THE CALCULUS

THE CALCULUS

A Genetic Approach

OTTO TOEPLITZ

New Foreword by David Bressoud

Published in Association with the Mathematical Association of America

The University of Chicago Press
Chicago · London

The present book is a translation, edited after the author's death by Gottfried Kothe and translated into English by Luise Lange. The German edition, *Die Entwicklung der Infinitesimalrechnung*, was published by Springer-Verlag.

The University of Chicago Press, Chicago 60637
The University of Chicago Press, Ltd., London
16 15 14 13 12 11 10 09 08 07   2 3 4 5

ISBN-13: 978-0-226-80668-6 (paper)

ISBN-10: 0-226-80668-5 (paper)

Library of Congress Cataloging-in-Publication Data

Toeplitz, Otto, 1881–1940.
    [Entwicklung der Infinitesimalrechnung. English]
    The calculus : a genetic approach / Otto Toeplitz ; with a new foreword by David M. Bressoud.
        p. cm.
    Includes bibliographical references and index.
    ISBN-13: 978-0-226-80668-6 (pbk. : alk. paper)
    ISBN-10: 0-226-80668-5 (pbk. : alk. paper)   1. Calculus.   2. Processes, Infinite.   I. Title.
    QA303.T6415 2007
    515—dc22
                                                                2006034201

♾ The paper used in this publication meets the minimum requirements of the American National Standard for Information Sciences—Permanence of Paper for Printed Library Materials, ANSI Z39.48-1992.

FOREWORD TO *THE CALCULUS:*

A GENETIC APPROACH BY OTTO TOEPLITZ

September 30, 2006

Otto Toeplitz is best known for his contributions to mathematics, but he was also an avid student of its history. He understood how useful this history could be in informing and shaping the pedagogy of mathematics. This book, the first part of an uncompleted manuscript, presents his vision of an historically informed pedagogy for the teaching of calculus. Though written in the 1930s, it has much to tell us today about how we might—even how we should—teach calculus.

We live in an age of a great democratization of calculus. A course once reserved for an elite few is now moving into the standard college preparatory curriculum. This began in the 1950s, but the movement has accelerated in the past few decades as knowledge of calculus has come to be viewed as a prerequisite for admission to the best colleges and universities, almost irrespective of the field that will be studied. The pressures and opportunities created by this popularization have resulted in two significant movements that have shaped our current calculus curriculum, the New Math of the 1950s and '60s, and the Calculus Reform movement of the 1980s and '90s. These movements took the curriculum in very different directions.

The New Math was created in response to the explosion in demand for scientists and engineers in the years following World War II. To prepare these students for advanced mathematics, the curriculum shifted to focus on abstraction and rigor. This is the period in which Riemann's definition of the integral entered the mainstream calculus curriculum, a curriculum that adopted many of the standards of rigor that had been developed in the nineteenth century as mathematicians extricated themselves from the morass of apparent contradictions revealed by the introduction of Fourier series.

One of the more reasoned responses to the New Math was a collective statement by Lipman Bers, Morris Kline, George Pólya, and Max Schiffer, consigned by many others, that was published in *The American Mathematical Monthly* in 1962.[1] In this letter, they called for the use of the "genetic method:" "The best way to guide the mental development of the individual is to let him retrace the mental development

---

[1] *The American Mathematical Monthly*, 1962, 69:189–93.

of the race—retrace its great lines, of course, and not the thousand errors of detail."
I cannot believe it was a coincidence that one year later the University of Chicago
Press published the first American edition of *The Calculus: A Genetic Approach.*

The Calculus Reform movement of the 1980s was born from the observation
that too many students were confused and overwhelmed by an approach to calculus that was still rooted in the rigor of the 1950s and '60s. In my experience, most
calculus students genuinely want to understand the subject. But as students encounter concepts that do not make sense to them and as they become confused, they
fall back on memorization. These students then emerge from the study of calculus
with nothing more than a capacity to handle its procedures and algorithms, with little awareness of its ideas or the range of its uses. In the 1980s, departments of mathematics were facing criticism from other departments, especially departments in en-
gineering, that we were failing too many of their students, and those we certified as
knowing calculus in fact had no idea how to apply its concepts in other classes. The
Calculus Reform movement tried to achieve two goals: to create student awareness
of and ability to work directly with the concepts of calculus, and to increase the accessibility of calculus, to make it easier for more students to learn what they would
need as they moved into subsequent coursework and careers. It created its own
backlash. The argument commonly given against its innovations was that it weak-
ened the teaching of calculus, but much of the resistance came from the fact that it
required more effort to teach calculus in ways that improve both accessibility and
understanding.

Today, the battles over how to teach calculus have receded. Most of the innova-
tive curricula created in the late 1980s have either disappeared or mutated into
something that looks suspiciously like the competition. The movement did change
what and how we teach: more opportunities for exploration, greater emphasis on
the interpretation of graphical and tabular information, a recognition that the ability to read and communicate mathematical ideas is something that must be developed, more varied and interesting problems, a recognition of when and how computing technology can aid in the transmission of ideas and insights. At the same
time, we still use Riemann's definition of the integral, and there is a lingering long-
ing for the rigor of epsilons and deltas. The problems that initiated both the New
Math and the Calculus Reform are still with us. We still have too few students prepared for the advanced mathematics that is needed for many of today's technical
fields. Too few of the students who attempt calculus will succeed in it. Too few of
those who complete the calculus sequence understand how to transfer this knowledge to other disciplines.

Toeplitz's *The Calculus: A Genetic Approach* is not a panacea now any more
than it was over forty years ago. But it brings back to the fore an approach that has
received too little attention: to look at the origins of the subject for pedagogical inspiration. As Alfred Putnam wrote in the Preface to the first American edition, this
is not a textbook. It is also not a history. Though Toeplitz knew the history, he is not
attempting to explain the historical development of calculus. What he has created is
a distillation of key concepts of calculus illustrated through many of the problems
by which they arose.

I agree with Putnam that there are three important audiences for this book, though I would no longer group them quite as he did in 1963. The first audience consists of students, especially those who are not challenged by the common calculus curriculum, who can turn to this book to supplement their learning of calculus. One of the penalties one pays for accessibility is a leveling of the curriculum. There are too few opportunities in our present calculus curricula for students with talent to wrestle with difficult ideas. I fear that we lose too many talented students to other disciplines because they are introduced to calculus too early in their academic careers and are never given the opportunity to explore its complexities. For the high school teacher who wonders what to do with a student who has "finished" calculus as a sophomore or junior, for the undergraduate director of mathematics who wonders what kind of a course to offer those students who enter college with credit for calculus but without the foundation for higher study that one would wish, this book offers at least a partial solution. Its ideas and problems are difficult and enticing. Those who have worked through it will emerge with a deep appreciation of the nature and power of calculus.

Putnam's second and third audiences, those who teach calculus and those preparing to teach secondary mathematics, have largely merged. Prospective secondary mathematics teachers must be prepared to teach calculus, which means they must have a depth of understanding that goes beyond the ability to pass a course of calculus. Toeplitz's book can provide that depth.

The final audience consists of those who write the texts and struggle to create meaningful curricula for the study of calculus. This book can help us break free of the current duality that limits our view of the calculus curriculum, the belief that calculus can be either rigorous or accessible, but not both. It suggests an alternate route through the historical development of difficult ideas, gradually building the pieces so that they make sense. Too many authors think they are using history when they insert potted accounts that attempt to personalize the topic under discussion. Real reliance on history should throw students into the midst of the confusion and exhilaration of the moment of discovery. I admit that in my own teaching I may celebrate confusion more than Toeplitz would have tolerated, but he does identify many of the key conceptual difficulties that once confronted mathematicians and today stymie our students. He conveys the historical role of conflicting understandings as well as the exhilaration of the discovery of solutions.

Mathematics consists of the abstraction of pattern, the overlay of abstracted patterns of different origin that exhibit points of similarity, and the extrapolation from the patterns that emerge from this overlay. The key to this process is a rich understanding of the conceptual patterns with which we work. The Calculus Reform mantra of "symbolic, graphical, numerical, and verbal" arose from the recognition that students need a broad view of the mathematics they learn if they are ever to be able to do mathematics. The beauty of Toeplitz's little book is that he forces precisely this broadening of one's view of calculus.

The first chapter is an historical exploration of the concept of limit. While it culminates in the epsilon definition, Toeplitz is careful to lay the groundwork, explaining the Greek "method of exhaustion" and the role of the principle of continuity.

Most significantly, he takes great care to explain our understanding of the real numbers and how it came to be. He demonstrates how limits and the structure of the real numbers are intimately bound together. Epsilons and deltas are not handed down from above with a collection of arcane rules. Rather, they emerge after considerable work on the concept of limit, appearing as a convenient shorthand for some very deep ideas. This is one of the clearest examples of the fallacy of a dichotomy between rigor and accessibility. That dichotomy only exists when we decide to "do" limits in one or two classes. Toeplitz's approach suggests a fundamental rethinking of the topic of limits in calculus. If it is an important concept, then devote to it the weeks it will take for students to develop a true understanding. If you cannot afford that time, then maybe for this course it is not as important as you thought it was. Though it would have been heresy to me earlier in my career, I have come to the conclusion that most students of calculus are best served by avoiding any discussion of limits.[2] It is the students who have a good understanding of the methods and uses of calculus who are ready to learn about limits, and they need a treatment such as Toeplitz provides.

Chapter 2 moves on to the general problem of area and the definition of the definite integral. I especially enjoy Toeplitz's brief section on the dangers of infinitesimals. I find it refreshing that Toeplitz completely ignores the Riemann integral which was, after all, created for the investigation of functions that do not and should not arise in a first year of calculus. Instead Toeplitz relies on what might be called the Cauchy integral, taking limits of what today are commonly called left- and right-hand Riemann sums. I agree with Toeplitz that this is the correct integral definition to be used in the first year of calculus.

The fundamental theorem of calculus provides the theme for the third chapter, which is the longest and richest. There is an extended section on Napier's tables of logarithms. Few students today are aware of such tables or the role they once played. But the mathematics is beautiful. Understanding the application of these tables and the complexity of their construction provides insight into exponential and logarithmic functions. Today's texts present these functions in the context of exponential growth and decay. While that is their most important application, it provides only a limited view of functions that are central to so much of mathematics. This chapter includes a discussion of the development of the relationship of distance, velocity, and acceleration. The difficulties Galileo encountered in conceiving, formulating, and then convincing others of these relationships often is underappreciated. Toeplitz pays Galileo his rightful due.

My favorite part of Chapter 3 is Toeplitz's discussion in the last section, "Limitations of Explicit Integration." He clarifies a point which, when ignored, leads to confusion among our students. That is the distinction between what he calls "com-

---

[2]For an illustration of how this can be done, see the classic text *Calculus Made Easy* by Sylvanus P. Thompson. It is a commentary on the hold limits have on our current curriculum that the most recent edition, St. Martin's Press, New York, 1998, includes additional chapters by Martin Gardner, one of them on limits.

putational functions," the standard repertoire built from roots, exponentials, sines, tangents, and logarithms, functions for which we can compute values to any pre-assigned accuracy, and the "geometrical functions," those represented by a graph of a continuous, smooth curve. He rightly points out that the challenge issued by Fourier series was to leave the limitations of computational functions and embrace the varied possibilities of geometrical functions. As Toeplitz says in the concluding lines of this chapter,

> Today's researchers have them both at their disposal. They use them separately or in mutual interpenetration. For the student, however, it is difficult to keep them apart; the textbooks he studies do not give him enough help, because they tend to blur rather than to sharpen the difference.

The emphasis on graphical representation that received impetus from the Calculus Reform movement has helped to promote student awareness of this distinction, but the two understandings of function still draw too little direct attention. I have found it helpful in my own classes to emphasize this distinction. For example, too many of my students enter my classes only knowing concavity as an abstract property determined by checking the sign of the second derivative. They are amazed to see that it can be used to describe geometric functions that are not given by any formula.

Finally, we come to Chapter 4 in which Toeplitz demonstrates how calculus enabled the solution of the great scientific problem of the seventeenth century, the explanation of how it is that we sit on a ball revolving at 1,000 miles per hour as it hurtles through space at speeds, relative to our sun, of over 65,000 miles per hour, yet we feel no sense of motion. Newton's *Principia* is a masterpiece. I teach an occasional course on it, and wish that all calculus students, especially those who are preparing to teach, would learn to appreciate what Newton accomplished. Toeplitz gives us an excellent if brief overview of Newton's work. He also explores the study of the pendulum, a remarkably rich source of mathematical inspiration.

There is much that all of us can learn about the teaching of calculus from this book, but I do not want to freight it with too much gravity. It is, above all, a delightful and entertaining introduction to mathematical problems that have inspired the creation of calculus. Read it for the sheer enjoyment of well-crafted explanations. Read it to learn something new. Read it to see classic problems in a rich context. But then take some time to ponder its lessons for how we teach calculus.[3]

DAVID M. BRESSOUD
Macalester College
St. Paul, Minnesota

---

[3]With thanks to Paul Zorn for his comments on a draft of this Foreword.

PREFACE TO THE GERMAN EDITION

In a paper presented before the Mathematische Reichsverband at Düsseldorf in 1926,* Otto Toeplitz outlined his ideas about a new method designed to overcome the difficulties generally encountered in courses on infinitesimal calculus. He called it the "genetic" method. "Regarding all these basic topics in infinitesimal calculus which we teach today as canonical requisites, e.g., mean-value theorem, Taylor series, the concept of convergence, the definite integral, and the differential quotient itself, the question is never raised 'Why so?' or 'How does one arrive at them?' Yet all these matters must at one time have been goals of an urgent quest, answers to burning questions, at the time, namely, when they were created. If we were to go back to the origins of these ideas, they would lose that dead appearance of cut-and-dried facts and instead take on fresh and vibrant life again."

To the young student interested in the exciting and beautiful aspects of mathematics, Toeplitz seeks to present the great discoveries in all their drama, to let him witness the origins of the problems, concepts, and facts. But he does not want to have his method labeled "historical." "The historian—the mathematical historian as well—must record all that has been, whether good or bad. I, on the contrary, want to select and utilize from mathematical history only the origins of those ideas which came to prove their value. Nothing, indeed, is further from me than to give a course on the history of infinitesimal calculus. I myself, as a student, made my escape from a course of that kind. It is not history for its own sake in which I am interested, but the genesis, at its cardinal points, of problems, facts, and proofs."

Toeplitz is convinced that the genetic approach is best suited to build the bridge between the level of mathematics taught in secondary schools and that of college courses. He also intends to lead the beginner in the course of two semesters to a full understanding and command of epsilontic, but he wants to advance him to the mastery of this technique only "gradually through gentle ascending." "The genetic method is the safest guide to this gentle ascent, which otherwise is

* Published in *Jahresbericht der deutschen mathematischen Vereinigung*, XXXVI (1927), 88–100.

not always easy to find. Follow the genetic course, which is the way man has gone in his understanding of mathematics, and you will see that humanity did ascend gradually from the simple to the complex. Occasional explosive great developments can usually be taken as indicators of preceding methodical progress. Didactical methods can thus benefit immeasurably from the study of history."

In the paper referred to above, Toeplitz announced that he hoped to present his method in the form of a textbook. He worked on it for many years, pursuing intensive historical studies of the development of infinitesimal calculus. In his lectures he constantly tried out new approaches, discussing the several parts with his students and searching always for new formulations.

He did not live to finish the book. He died in Jerusalem on February 19, 1940, after years of deep mental suffering. His decision to emigrate was made only at the last moment; he left Germany early in 1939. In those years he rarely found the strength for intensive scientific work.

The manuscript of the present volume was found among his papers. It covers the genesis of mathematics up to Newton and Leibniz and was intended as a textbook for the first semester of the course. Marginal notes indicate that Toeplitz was planning to revise the last parts dealing with applications to mechanics, but I felt that I should present the work as I found it in manuscript and not make any changes except for some necessary editing. For the benefit of readers interested in the historical aspects, I added a chronological table and such references to the literature as seemed important for the historical arguments.

The appended exercises, which were chosen by Toeplitz himself, must be regarded as indispensable for the purposes of the book. With the exception of a few, which served to round out the text, all the problems were used and tried out by Toeplitz in his discussion periods. (These exercises, which were expected to be worked out very carefully, are in some cases but loosely connected with the text.) They relieve the lectures of occasional detail which might interfere with the main line of development, but they do contain some important supplementary material. Toeplitz always considered discussion periods most important for the discovery of students with superior mathematical ability. For his ideas on this timely subject we may give at least the reference below.*

The original title of the manuscript was "Introduction to Infinitesimal Calculus, Vol. I." It seemed to me, however, that the specific character of the book, and its unique position among other introductory works, called for a more descriptive title. Its genetic method provides such deep understanding of the basic ideas as cannot be achieved through a systematic presentation, and, beyond this, it achieves such a thoroughly balanced view of the development of infinitesimal calculus along its principal lines that I believe the title chosen for the book does more justice to its nature. (The material earmarked for the second volume of the work may be insufficient to permit its publication in even rough

---

* "Die Spannungen zwischen den Aufgaben und Zielen der Mathematik an der Hochschule und an den höheren Schulen," *Schriften des D.A.M.N.U.*, No. 10 (Leipzig, 1928), pp. 1–16.

approximation to the author's intention. But the present volume really covers the principal points of the development.)

I express my thanks to Mrs. Erna Toeplitz, Jerusalem, and her assistants for the copies they made of the original manuscript; also, for help in making corrections and for valuable suggestions, to Professors J. J. Burckhardt, Zurich; J. O. Fleckenstein, Basel; H. Ulm, Munster; K. Vogel, Munich; and to my colleagues at Mainz University, H. E. Dankert, Dr. H. Muller, Dr. W. Neumer, and Professor H. Wielandt. The diagrams were made by Dr. W. Uhl, Giessen.

G. Köthe

Mainz
Easter 1949

PREFACE TO THE ENGLISH EDITION

This is not a textbook in the calculus, nor is it a history of the calculus. Over a period of many years Toeplitz sought in his teaching to evolve what he described as a "genetic approach" to the subject. He aimed for nothing less than a presentation of the calculus that would do justice to the growth and development, in the course of time, of its central ideas. The choice of topics is that of a mathematician concerned for what have proved to be crucial concepts and techniques of the calculus. The organization and exposition are determined by the historical evolution of these themes from their beginnings with the Greeks down to the present. Toeplitz was convinced that only through this genetic approach could students come to a full understanding of the significance of the concepts and techniques which constitute the calculus. To be sure, he was addressing himself to a German audience with the mathematical training provided by the Gymnasium, but what he has offered in this volume is equally relevant for serious students and teachers of the calculus in America.

There are three groups to whom this volume is particularly directed. First are the students in a standard course in the calculus. This book is no substitute for a text, but it should be a most valuable supplement for those students who seek to know how the calculus arose and how it has come to its present form. Second are the teachers of the calculus. For them to have an understanding of the origins of the subject and the development of its concepts must be a matter of professional concern. Third are those preparing to be teachers of mathematics. They will already have studied the calculus, but they have an obligation today as never before to secure a thorough grounding in all the principal branches of their subject. To all of these, and to those who simply seek to know what the calculus is really all about, this genetic approach is offered.

Alfred L. Putnam

TABLE OF CONTENTS

IV. APPLICATIONS TO PROBLEMS OF MOTION   133

EXERCISES   173

BIBLIOGRAPHY   183

BIBLIOGRAPHICAL NOTES   185

INDEX   191

1

THE NATURE
OF THE INFINITE PROCESS

Two subjects, analytic geometry and differential and integral calculus (or infinitesimal calculus, as it is more comprehensively and appropriately called), form the principal courses for beginners in our university program of mathematical instruction. The distinction between the two subjects seems clear enough from their names alone; the one deals with geometry, the other with calculation. In reality, however, the principle which this distinction expresses is not sound; infinitesimal processes tie in as well with geometric objects as with calculational ones, and the geometry of figures of the second and third degree in the plane or in space can culminate in a purely calculational determinant theory. Therefore, the true distinction between these two subjects is that infinitesimal calculus uses infinite processes, whereas analytic geometry avoids them. That distinction extends, far beyond the beginning courses, throughout mathematics and offers the only serious basis from which to proceed to a classification of the whole science. As mathematics develops, that distinction becomes ever clearer, while that between geometry and calculation fades away. Our first task, therefore, will be to delineate the essential nature of the infinite process. In later chapters we will consider such particular kinds of infinite processes as differentiating, integrating, and summing infinite series.

1. THE BEGINNINGS OF GREEK SPECULATION ON INFINITESIMALS

It bespeaks the true greatness of a new idea if it appears absurd to contemporaries encountering it for the first time. The paradoxes of Zeno are the first indication to us of the emerging idea of the infinite process at a time about whose intellectual life we otherwise know but little. To their author they were doubtless not the quasi-punning puzzles which are reported to us, an impression which is accentuated by the form given to the argument. In the report to which we largely owe our knowledge of the paradoxes, Aristotle certainly discounts their disguise by arguing: "I cannot go from here to the wall! To do so, I would first have to cover half the distance, and then half the remaining distance, and then again half of what still remains; this process can always be carried on and can never be brought to an end." It is unreasonable to suppose that Zeno was unaware that the times needed to traverse these successive halves themselves become shorter and shorter.[1] * He is protesting only against the antinomy of the infinite process which wc encounter in proceeding along a continuum. And this protest, expressed with youthful exuberance but recorded almost against his will, indicates that mathematicians had then first dared to undertake the summation of infinitely many, but ever decreasing, bits of time, like

$$1 + \tfrac{1}{2} + \tfrac{1}{4} + \tfrac{1}{8} + \ldots .$$

It is interesting to compare this report of Aristotle's with one of the fragments, likewise from the fifth century b.c. (which has come down from Anaxagoras): "There is no smallest among the small and no largest among the large; but always something still smaller and something still larger." These words seem trivial to us today; they certainly were not so in the Age of Atomism. This was not the atomism of which we were thinking in imagining discrete material atoms distributed in space but a theory which anticipated discontinuity of space itself and considered the possibility that a line segment might not be indefinitely divisible. Zeno's paradoxes go far beyond a flat rejection of this atomism as expressed in the statement of Anaxagoras.[2] Though we do not know much about the mutual relations of Eleatics, Pythagoreans, and other philosophical schools, we cannot doubt that Zeno's criticism is directed against some first, uncertain claims of a new mathematics trying to replace the naïve atomistic view based on intuition with laws found by systematic reasoning.

The conflict to which Zeno's paradoxes give expression comes to the open with the "Pythagoreans" at the moment when they discover the "irrational," thereby facilitating the rise of the idea of the infinite process and laying its base, which is valid to this day. Just what is this "irrational"? It is contained in the discovery of a side and the diagonal of one and the same square being "incommensurable," that is, lacking a common measure.

The "carpenter" rule for constructing a right angle had long been known: Make the two sides of a triangle 3 and 4 ells long and incline them toward each other in such a way that the line segment connecting the two ends measures exactly 5 ells; then the angle is a right angle, and the triangle a right triangle (Fig. 1). Mathematicians kept trying to find an analogous whole-number relationship among the sides for the much more obvious case of a right triangle with equal sides (Fig. 2). They divided each leg into five equal parts and then laid off one of these parts on the hypotenuse. This part seemed to be contained in the hypotenuse seven times, yet not quite exactly; the hypotenuse was a bit too long. They tried the same with twelve divisions of the legs; the hypotenuse this time fitted with seventeen such parts much more exactly than before but still not entirely. All such efforts to discover a "common measure" for both segments—legs and hypotenuse—were fruitless; finally, it was recognized that this quest must remain vain—that there exists no such common measure.

* [See Bibliographical Notes, pp. 185–89, below.]

There are two proofs of this impossibility. The first proof is based on easy observations about even and odd numbers:

1. The square of an even number is always divisible by 4:

$$(2n)^2 = 4n^2 .$$

2. The square of an odd number is always odd:

$$(2n + 1)^2 = 4n^2 + 4n + 1 = 2(2n^2 + 2n) + 1 .$$

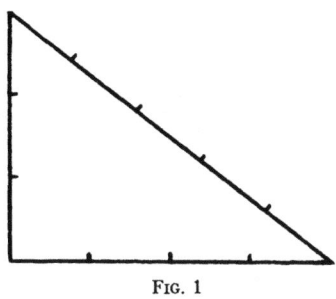

FIG. 1

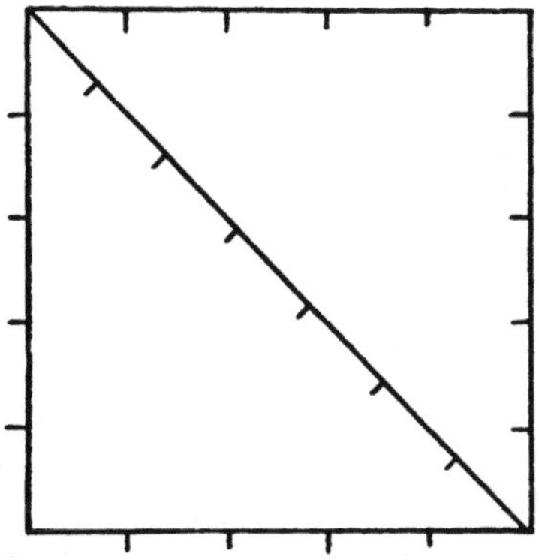

FIG. 2

From these two observations follow two more:

3. If the square of a number is even, then the square is divisible by 4.
4. If the square of a number is even, then the number itself is even.

The impossibility proof itself is indirect. Suppose the side a and diagonal d of a square had a common measure, e, and $d = pe$ and $a = qe$. By the Pythagorean theorem

$$d^2 = a^2 + a^2 = 2a^2 ;$$

therefore,

$$(pe)^2 = 2(qe)^2 ,$$

so that

$$p^2 = 2q^2 . \qquad (1.1)$$

Here it may be assumed that p and q have no common factor except 1, since otherwise the common measure e would have been chosen too small and could be so enlarged.

Since the right side of equation (1.1) is evidently even, the left side is too. According to Observation 4, p must then be even. On the other hand, since p^2 is even, it is divisible by 4 according to Observation 3. But then the right side of equation (1.1) is divisible by 4, so that q^2 must be divisible by 2 and therefore even. Once more application of Observation 4 shows q itself even. Therefore, p and q would both be even, contrary to the express assumption that p and q

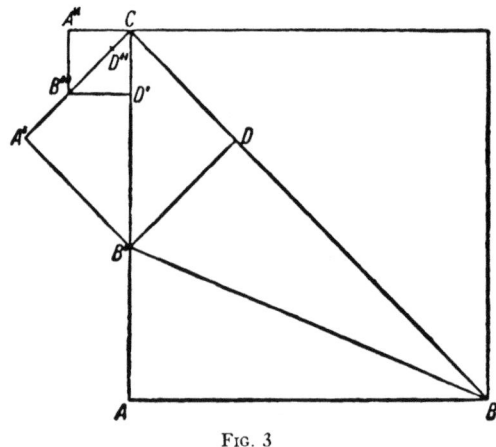

Fɪɢ. 3

have no common factor except 1. The initial supposition of the proof—that the side and diagonal of the square had a common measure—has led to a contradiction and is, consequently, disproved.

The second impossibility proof uses an elementary geometrical consideration instead of facts about even and odd numbers: In the square in question (Fig. 3), lay off on the diagonal beginning at B a segment BD of the same length as side $AB;$ at D erect a perpendicular meeting side AC in $B';$ join B' and B. Triangles ABB' and DBB' are congruent, since two pairs of corresponding sides are equal and the angles opposite the larger side are equal; therefore, $AB' = DB'$. The angle ACB is half a right angle; therefore, $B'CD$ is an isosceles right triangle, and $DB' = DC$. It has been established that

$$AB' = B'D = DC . \qquad (1.2)$$

Now erect at C the perpendicular to CD and draw through B' the parallel to CD which meets that perpendicular at A'. A square $A'B'CD$ is obtained which is smaller than the original, $ABCD$, since the diagonal $B'C$ is already covered by one of the sides of the original. To this new square there is now to be applied the same procedure that was applied to the original; mark off a segment $B'D'$ on the diagonal equal to the length of side $A'B'$ and at D' erect the perpendicular to the diagonal meeting side $A'C$ in B''. Then, as before,

$$A'B'' = B''D' = D'C. \qquad (1.3)$$

It is clear that the procedure continues indefinitely and never terminates; instead, each time there remains a piece of the diagonal smaller than the previous remainder

$$CD > CD' > CD'' > CD''' > \ldots . \qquad (1.4)$$

Each of these remainders is the difference of the diagonal and side of one of the successive squares:

$$CD = CB - AB, \qquad CD' = CB' - A'B',$$

$$CD'' = CB'' - A''B'', \ldots . \qquad (1.5)$$

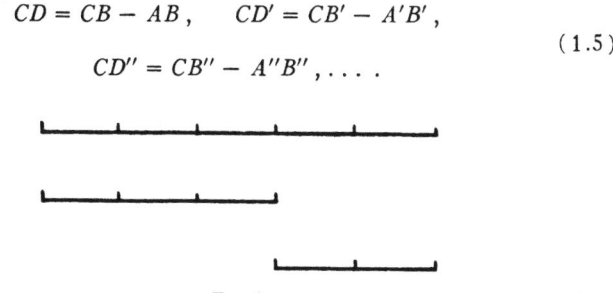

Fig. 4

This elementary geometrical consideration is the necessary preliminary to the proof; the proof itself is indirect. Suppose the side and diagonal of the square to be commensurable; that is, suppose that there is a common measure of the two—an interval E a certain exact multiple of which would equal the side of the square and a certain other exact multiple of which would equal the diagonal. Then it is only necessary to observe (Fig. 4) that the difference of any two intervals which are both exact multiples of E is likewise an exact multiple of E. So, if CB and AB are exact multiples of E, then from equation (1.5) CD is too. And as $A'B' = CD$, $A'B'$ is then an exact multiple of E. The diagonal CB' of the square $A'B'CD$ is such that $CB' = CA - AB' = AB - CD$—this last by equation (1.2)—and CB', being the difference of two exact multiples of E, is therefore an exact multiple of E. This property having been proved for the side and diagonal of the square $A'B'CD$, it follows by the same kind of argument for all later squares.

The indirect proof can now be completed by arguing to a contradiction. The intervals appearing in (1.4), on the supposition that there is a common measure for the side and diagonal of the original square $ABCD$, must all be exact multiples of E. On the other hand, equation (1.4) asserts that the multiples of E

continually decrease without either terminating or ever becoming zero. This is impossible for multiples of a fixed interval, since, if the initial term were 1,000 times E, then CD' would be a smaller exact multiple of E—at most 999 times E. At the very latest the 1,000th member in this chain would have to be smaller than E and yet still a multiple of E, and, therefore, zero times E, contrary to what has already been proved. This is the contradiction to which we have been led by the supposition that there is a common measure of the side and diagonal of a square; the supposition is therefore untenable.

Complete darkness covers the origins of this first impossibility proof. This great discovery, more than anything else, inaugurated the character of modern mathematics. The oldest and least ambiguous evidences are found in Plato and Aristotle.[3] The latter, repeatedly referring to the subject, alludes to the first-mentioned proof, which later appears again with Euclid. Plato puts considerable emphasis on the fundamental nature of this discovery. In the *Laws*, at the point where he assigns that mathematical discovery a place in higher school instruction, he mentions that he first learned of it when he was a comparatively old man and that he had felt ashamed, for himself and for all Greeks, of this ignorance which "befits more the level of swine than of men." Especially in the dialogue dedicated to the memory of one of the greatest Greek mathematicians, Theaetetus,[4] who had just fallen in battle, he gives an account of these matters. There he tells how Theodorus, the teacher of Theaetetus, a well-known Sicilian mathematician who was born about 430 B.C., had lectured to his students on the proof that the side of a square of area 3 square feet is incommensurable with an interval 1 foot long, and similarly for squares of area 5 up to 17 square feet, 9 and 16 square feet excepted, of course. From this citation it is quite clear that the teacher Theodorus was already in possession of a well-developed theory of such facts, the case of 2 square feet not even being mentioned, although Plato contrasts him with his student Theaetetus, who introduced a more abstract and general approach to this theory.

The essential content of Greek mathematics is found in the *Elements* of Euclid,[5] about 300 B.C., and in the writings of mathematicians and commentators who came after him; but there is only the content, not the history of its development. Nothing but such fragments as those cited above permits us a fleeting glimpse into the beginnings of this mathematics, which is by no means the work of Euclid. One of the most important and comprehensive of these fragments consists of a few pages from what is probably the oldest textbook of Greek mathematics, or at least from the hand of Hippocrates,[6] who lived about 450 B.C. It shows us the next stage in the development of the theory of infinite processes. Hippocrates halves a circle (Fig. 5) by diameter AB; then, with the midpoint of the lower semicircle as center, he draws a circle passing through A and B. He asserts that the crescent (shaded in Fig. 5) has the same area as the square on the radius BM of the original circle.

The proof rests on a lemma, the basis of which unfortunately is not given in the extant pages. It claims: The areas of two circles are to each other as the squares of their radii (Fig. 6). From this it follows that segments of two different circles which subtend equal angles (Fig. 7) are to each other as the squares of the radii. This is proved first for angles which are an aliquot part* of the full angle and later for arbitrary angles. Next Hippocrates connected the midpoint D of the upper semicircle (Fig. 8) with A and B and found that these lines touched the large circle at A and B without entering into it. The crescent under consideration consists, therefore, of three areas designated in the figure as α, β, and γ. Area α is a segment of the given circle subtending one-fourth of the full angle; δ is a segment of the large circle subtending one-fourth of the full angle. In accordance with the above lemma, α and δ are therefore to each other as the squares of the radii of the two circles, that is, as $AM^2 : AD^2$, which is evi-

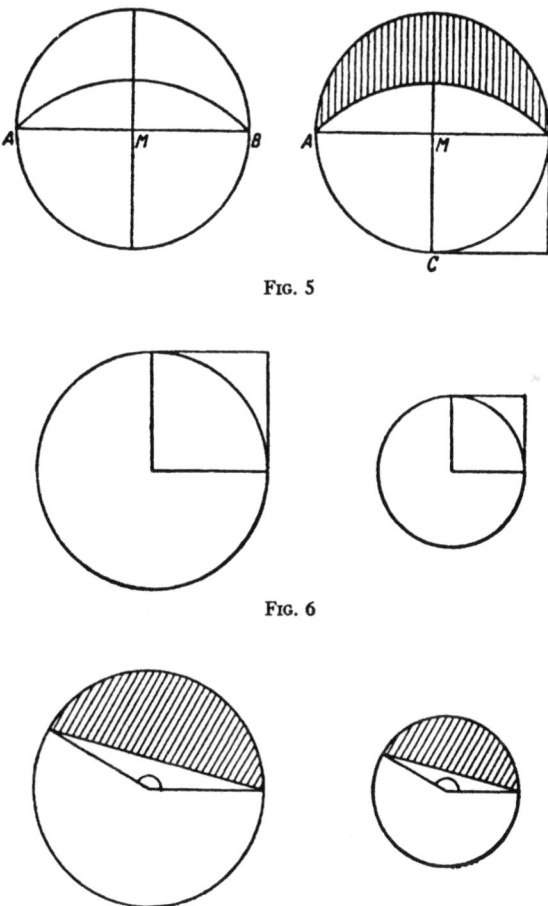

FIG. 5

FIG. 6

FIG. 7

*an integer divisor

dently 1:2. Hence α and likewise β are half of δ, which means $\alpha + \beta = \delta$, and the area of the crescent $\alpha + \beta + \gamma = \delta + \gamma =$ area of triangle $ABD = BM^2$.

The fundamental importance of this discovery lies in proving the possibility of areas bounded by curved lines being commensurable with areas bounded by straight lines. The problem of "squaring the circle" derived its great appeal from this discovery. For to accomplish for the very simplest curvilinear figure what had been accomplished for another curvilinear figure was surely a powerful challenge which attracted mathematicians for two millenniums; and even today, after the impossibility of its solution has long been demonstrated, nonmathematicians who do not understand this impossibility proof are still trying to "square the circle." Hippocrates clearly understood the problem and pursued his goal in a most methodical fashion. He tried to find other meniscuses (crescents, or lunulae, such as the changing phases of the moon present in all possible forms) which would have the same property as the first one, in order to build up eventually the whole circle from such crescents. He found two additional ones, very cleverly designed, which, while not commensurable to the

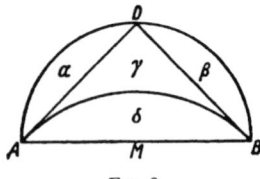

FIG. 8

square over the chord, had areas equal to that of a certain polygon. Transforming such figures with compass and ruler into a square of equal area—to "square" it—was a problem which the mathematicians of those days apparently were already mastering completely. The problem of squaring the circle in this manner had thus been well defined, but this first, broad attempt to solve it ended in failure because the circle could not be built up from the crescents which Hippocrates had constructed.

In those days, too, there were people who missed the point of the problem, for example, the Sophist Antiphon, who lived in Athens at about the same time as Hippocrates. Aristotle relates of him that he inscribed a square in a circle (Fig. 9), then constructed isosceles triangles over its sides forming a regular octagon inscribed in the circle, then similarly a regular polygon of sixteen sides, and so on. As any such rectilinear polygon could be transformed into a square, Antiphon believed that there must be a polygon of a sufficiently great number of sides which would be identical with the circle and that the square into which that polygon could then be transformed would be the solution.

This argument, which was valid enough for a concrete circle drawn with even a fine stylus, was rejected by Aristotle as invalid for the ideal circle of geometry. Hippocrates, too, entertained this clear concept of an exact geometry dealing with ideal configurations, as is shown by the mentioned citations from Anaxagoras. Here, where the argument concerns a concrete special subject, we see more clearly than by the study of a philosophical debate how the thinking of the Sophists differed from the true scientific thinking born in those days. Today as well, after two thousand years of history of science, there are still many who are unable to comprehend the ideal nature of the objects of mathematics or who regard it as of minor importance or as a more or less superfluous "finer point." Thus the Sophists appear here not as ridiculous fellows but as representatives of a mental attitude which exists to this day and continues fighting pure science. Today's quarrels between the "applied" and "pure" mathematicians are but the continuation of that age-old struggle. It seemed of special importance to point this out right at the beginning of our course.

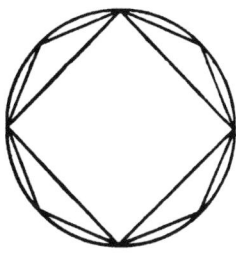

FIG. 9

2. THE GREEK THEORY OF PROPORTIONS

The discovery of the irrational—of incommensurable segments—meant a revolution of the whole of geometry. Let us make this clear by an example, namely, by the theorem that the areas of two triangles with equal altitudes (Fig. 10) are to each other as their bases, $A:B = a:b$.

For the case $a = b$ the theorem had long been proved that triangles of equal base and equal altitude have equal areas. From this, the theorem is readily proved also for the case of a and b being commensurable (e.g., $a:b = 3:2$); if we place next to A another triangle of base a, and next to B two other triangles of base b (Fig. 11), the two resulting large triangles have equal bases (for $a:b = 3:2$ means nothing else than $2a = 3b$) and hence equal areas; that is, $2A = 3B$, or also $A:B = 3:2$.

With the discovery of incommensurable segments, however, this proof, and with it all proofs of geometrical theorems concerning proportions (i.e., the whole theory of similar figures), became questionable. In fact, the very definition of proportionality was shaken, and mathematicians had to reconsider what was even meant by the statement that, for example, the areas of two triangles A and B "are to each other" as the bases a and b.

We do not know exactly when this crisis occurred or who overcame it. But in Book v of Euclid's *Elements* we find the finished edifice of the theory of proportions which from then on formed the basis of Greek mathematics. Even today, although transformed, it is still the decisive feature of our number concept and theory of infinite processes. We shall therefore briefly report on this chapter of Euclid's.

He begins with the following two ingenious definitions designed to circumvent the difficulty presented by the incommensurable:

1. If for every two natural numbers p and q the three relations $qa < pb$, $qa = pb$, and $qa > pb$ imply, respectively, $qA < pB$, $qA = pB$, and $qA > pB$, we say that $a:b = A:B$.

2. If there is a single pair of natural numbers p_0, q_0 for which $q_0a < p_0b$, while $q_0A > p_0B$, we say that $a:b < A:B$, or $A:B > a:b$.

In the case of commensurable quantities, Definition 1 comprises the old

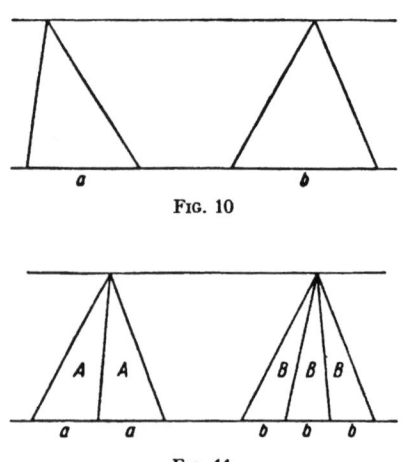

FIG. 10

FIG. 11

definition because the second of the three possibilities holds. Definition 1 applied to the above example of the triangles not only saves the proportion concept but also permits the reorganization of the proof. For, if p and q are any two natural numbers, and if q triangles of base a are placed side by side, and p triangles of base b, then one of three possibilities must be realized:

$$qa < pb, \qquad qa = pb, \qquad qa > pb.$$

Now we easily obtain the following lemma: If two triangles U and V have equal altitudes (Fig. 12) but unequal bases, with $u < v$, then $U < V$. For if we lay off the smaller segment u on v, there is over u a triangle equal in area to U which, however, is only a part of V and hence smaller in area than V. But if $qa < pb$, the areas of the two large triangles will be in the same ratio of inequality, that is, $qA < pB$, and so on.

The two definitions cited above, however, would not support the entire theory of proportions. As is well known, Euclid begins his planimetry—contained in his first four books and in its essence still taught in our schools—with a list of definitions, postulates, and axioms, that is, statements which are not proved but based on intuition. In developing next, in Book v, a general theory of quantity and proportions—of segments or areas or volumes, or time intervals, or whole numbers, or still others—he abrogated altogether any proof based, explicitly or tacitly, on graphic figures or on intuition. To an even higher degree than in his planimetry he forced himself to list completely all the principles on which he builds his further theory. And that he did. Aside from some axioms, such as "The whole is greater than any of its parts" (which was tacitly assumed in the above proof), or "Equals added to equals give equals," or similar ones, he listed one somehow less readily thought of, though evident enough. We call it the *axiom of continuity:*

> *If* A *and* B *are two quantities of the same kind, say, two segments or two time intervals, and if* A *is the smaller of the two, then it is always possible to find a multiple of* A, nA, *which is greater than* B.

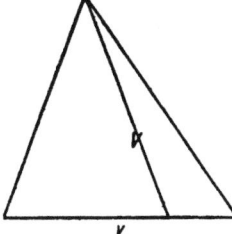

Fig. 12

This axiom, in combination with the two definitions cited above, is the foundation upon which Euclid erected the whole theory of proportions. The details of how this was done do not interest us here, the less so because later in this book we shall discuss similar reasoning in modern form. Of great importance for the genesis of the infinite process, however, is Euclid's other application of the continuity axiom—the *method of exhaustion.*

3. THE EXHAUSTION METHOD OF THE GREEKS

We shall also explain as an example, this type of Greek infinitesimal mathematics. As such we choose the first and simplest theorem which Euclid, reproducing the ideas of Eudoxus, used in Book xii. His methods became the model for the work of the great Archimedes and the fountainhead of modern analysis. We mentioned the lemma which Hippocrates used as his starting point to show that *the areas of two circles are to each other as the squares of their radii.* Unfortunately, we do not know how he arrived at this lemma. As in the case of the theory of proportion, we do not know whether the method which we call the *method of exhaustion* was already familiar to the mathematicians of that time or whether it was first invented by Eudoxus or by the other mathematicians of the Platonic Academy.

Euclid mentioned a proof based on the continuity axiom which was the same idea Antiphon had proposed for squaring the circle (see Fig. 9). If the contemporaries of Hippocrates had the knowledge of constructing with compass and ruler a square equal to any arbitrary polygon, they certainly must have known that the areas of two similar polygons inscribed in two circles are to each other as the squares on the radii of the two circles.* From this it follows that regular n-sided polygons which can be inscribed in the two circles of the lemma are to each other as the squares of the radii of these circles.

If the idea of Antiphon were mathematically tenable, that a regular polygon with sufficiently many sides is identical with the circle, the lemma would indeed be established by the reasoning given above. But as geometry treats of ideal lines, and not physical drawings, this reasoning is not satisfactory; a fuller proof is necessary. It must be shown that the lemma is not invalidated by the fact that there is always a difference between the circle and an inscribed n-sided polygon, no matter how large n be chosen. To overcome these and similar difficulties, the Greek mathematicians applied the continuity axiom. So, not satisfied with the vague and intuitive evidence of indefinite approximation, they introduced a clear logical proof, placing all evidence in this very axiom which had been found indispensable for another purpose.

The continuity axiom is here reformulated in the following way:

If initially $a > \epsilon$, and then diminished by at least half of itself, and the remainder again by at least half of itself, and so on, a point will be reached where the remainder is less than ϵ.

For, according to the continuity axiom, there exists a multiple $n\epsilon$ of ϵ which is greater than a. Now 2ϵ is double ϵ, and 3ϵ is less than twice this double, that is, less than $2^2\epsilon$; similarly, $4\epsilon < 2^3\epsilon$, and so on; hence

$$a < n\epsilon \leq 2^{n-1}\epsilon.$$

Thus ϵ may be doubled so often as to become $> a$, or, in other words, a can be halved in this manner, until the remainder is less than ϵ. And if, instead, it is diminished at each step by more than half, the desired result is achieved a fortiori.

This lemma permits us to estimate the error involved at every step of Antiphon's procedure. His inscribed square differs from the circle by the region shown shaded in Figure 13. In the square circumscribed about the circle shown in the figure, which is obtained from the inscribed square by placing four isosceles right triangles on its sides, all triangles are equal to each other, and equal also to the four triangles which compose the inscribed square. The shaded region is only part of the total of the four added triangles; since these are equal to the inscribed square, the shaded region is shown to be less than the inscribed

* [The phrase "inscribed in two circles" is missing in the German original.—Editor.]

square. Hence the difference between the inscribed square and the circle is less than half the area of the circle.*

In the case of an octagon, an isosceles triangle is placed on each side of the inscribed square, each triangle being half as large as the rectangular boxes built on the side (Fig. 14). The shaded area represents the difference between the inscribed regular octagon and the circle. It is less than half the rectangular boxes and hence less than the sum of the four added triangles by which the square was altered into the octagon. In other words, if the square is cut out of the circular disk, it loses more than half its area; if the four triangles, too, are eliminated (i.e., the whole octagon), the circle again loses more than half the remaining area. Continuing to the sixteen-sided polygon, we again cut away more than half the remainder, and this may be continued indefinitely.

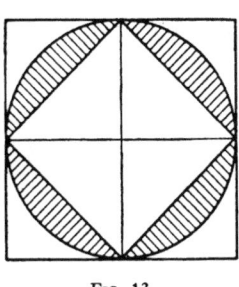

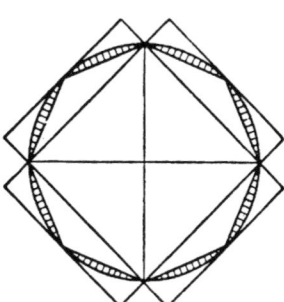

FIG. 13   FIG. 14

But the continuity axiom permits us to conclude that, if ϵ is the area of some preassigned small square, the continuation of the cutting process described above will at some time lead to a remainder whose area is less than ϵ. Let f be the area of the n-sided polygon for which $K - f < \epsilon$, where K represents the area of one of the two circles of the lemma. Let L be the area of the other circle and r and s the respective radii. We shall now have to prove $K:L = r^2:s^2$. If this were false, then K would have to be either larger or smaller than required by this proportion. If we assume the alternative—that K is too large for the proportion—then there would be a smaller area $K - \delta$ to satisfy the proportion

$$(K - \delta):L = r^2:s^2 .$$

Now we assign to the as yet undetermined little square ϵ, mentioned above, a value less than or equal to the excess area δ and, with the aid of the continuity

---

* [A modification of the argument in the last two sentences of the German original makes it clearer: The area of the inscribed square thus equals half that of the circumscribed square, and, hence, more than half the area of the circle. Wherefore the sum of the areas of the four shaded segments, which together are the difference between the circle and the inscribed square, is less than half the area of the circle.—EDITOR.]

axiom, determine that n-sided polygon whose area $K - f$ satisfies the inequality

$$K - f < \epsilon \leq \delta, \quad \text{or} \quad K - \delta < f.$$

Now let g be the area of the regular polygon of the same number of sides n inscribed in the other circle of area L; wherefore $g < L$, since g is only a part of L. Then we have both $f > K - \delta$ and $g < L$; hence

$$f:g > (K - \delta):L = r^2:s^2.$$

On the other hand, the proportion

$$f:g = r^2:s^2$$

holds for polygons, as noted initially. Hence our indirect proof has shown that the assumption

$$K:L > r^2:s^2,$$

leads to a contradiction. The assumption $K:L < r^2:s^2$ would be contradictory, too, if merely the two circles, whose roles are intrinsically the same, were interchanged.

Thus we have given a logically complete indirect proof. The mystery which otherwise attaches to the infinite process is absorbed by the continuity axiom. Accepting this axiom once and for all, we have no need to resort in every proof to a new vague intuitional evidence. This is the meaning of the Greek *method of exhaustion*.

4. THE MODERN NUMBER CONCEPT

The discussion of the infinite processes of the Greeks required the outlining of the Greek theory of proportions, which in turn led us to the continuity axiom, the core of the methods involving infinite processes. Similarly, a discussion of modern infinite processes requires an awareness of the number concept, which, patently or latently, has formed the basis of infinitesimal mathematics since its beginnings. The question of what is a number is hardly touched on in school mathematics. After the beginner has unquestioningly accepted the operating with natural numbers, fractions, and negative numbers, the practice of measuring leads him next to the infinite decimal fraction in terms of which the measure of any segment or area can be given with unlimited accuracy. Thereafter, in the upper grades, the idea of the infinite decimal fraction is the number concept with which we operate.

This number concept held also in modern mathematics from Descartes and Kepler on down to that moment in the nineteenth century when mathematicians began to reflect seriously on the nature of numbers. For our present inquiry, which attempts to trace genetically the development of the various infinite processes up to the time of Kepler, the definition of a number as an infinite decimal fraction will therefore be acceptable until the moment when our study of problems and methods has been carried to the threshold of the nineteenth century, and the difficulties have been opened up which demanded a revision of basic concepts. We shall find that this development arises from the same scruples which the Greeks had felt in laying the foundation for their theory of proportions and of which the polemic of Aristotle against his admired teacher Plato, difficult to interpret though it is, gives us an approximate idea.[7]

So far, however, we look in the opposite direction, toward the origin of the concept of number as an infinite decimal fraction. It is around 1600 that we find that concept in the form and usage to which we are accustomed. Sine tables were then systematically transformed into decimal notation—by Viète in 1579 and by Stevin in 1585—with the invention of the decimal point.[8] The inventors of logarithms soon cast these into the form of decimal fractions. The discoverers of the infinitesimal calculus tacitly or subconsciously adhered to the number concept throughout the seventeenth century.

Did it really originate around 1600? Was it invented by Viète and Stevin? To answer this question, we must see clearly to what extent the number concept envisages a decimal fraction.

The object to which the mathematicians of that time had to apply the number concept were the sine tables. In those times it was the needs of astronomy which directed the progress of mathematics. The great improvement of astronomical instruments achieved by Tycho Brahe and the refinement of astronomical measurements due to the invention of the telescope called for the elaboration of mathematical tables of higher accuracy and of suitable computational methods.

Let us compare the decimal tables of the early seventeenth century with the tables of the late sixteenth century, that is, with the great *Opus palatinum* begun by Rhäticus in 1550 with considerable public funds appropriated for the purpose and continued by Pitiscus.[9] These tables did not yet use decimal fractions. This, however, was in fact a very minor difference—one merely of form. Indeed, it was nothing but a slight change in form when, in 1660, decimal fractions and the decimal point were adopted. The essence of the vast mathematical practice was not changed at all. For Rhäticus already had taken the radius of the basic circle, not equal to unity, but equal to 10^{10} times greater than those in our sine tables. Hence we can hardly say that this number concept was essentially different from that of his successors. The actual calculations he carried out were exactly the same; all differences were quite minor.

Let us now go back a few more centuries, to 1250, when the Western world began to awaken from a long period of stagnation. Here we find—to give just one sample—Leonardo Pisano[10] solving the cubic equation

$$x^3 + 2x^2 + 10x = 20$$

by

$$x = 1^p 22' 7'' 42''' 33^{IV} 4^V 40^{VI} .$$

This is a sexagesimal fraction, which means

$$x = 1 + \frac{22}{60} + \frac{7}{60^2} + \frac{42}{60^3} + \frac{33}{60^4} + \frac{4}{60^5} + \frac{40}{60^6},$$

in the same way as $x = 1 \cdot 47$ means

$$x = 1 + \frac{4}{10} + \frac{7}{100}.$$

How Pisano found this solution we are not told. Calculated by modern methods and then put in sexagesimal form,

$$x = 1^p 22' 7'' 42''' 33^{VI} 4^V 38.5^{VI}.$$

This reveals an amazing mathematical skill on the part of Pisano, a complete mastery of computations with sexagesimal fractions, addition, multiplication, extraction of square roots, and so on. And it is computation which, after all, constitutes the core of the modern number concept. Whether 10 or 60 is taken as the base is quite irrelevant. While it is mistaken to see this number concept emerging already at the moment when an angle was first sexagesimally subdivided but when all the above-mentioned arithmetical operations were yet unknown, it is likewise mistaken to attach importance to the external form under which the operations are carried out.

Jordanus Nemorarius[11] and Leonardo Pisano did not invent these operations; they adopted them, completely developed, from the Arabs. Oswald Spengler[12] advanced the interesting thesis that the modern number concept is essentially different from that of the Greeks; that the system of mathematics built on the one and the other number concept are intrinsically different. To which degree that thesis is tenable will readily appear from this history of infinitesimal calculus. The very posing of this question by Spengler certainly is thought-provoking. If, however, we regard his raising of the question as the truly valuable part of his statement, we may wonder why he, being so attentive to Arabic culture, failed to raise a similar question with respect to the thinking of the Arabs. Indeed, Arabic mathematics deserves the special interest of anyone who tries to understand the difference between the Greek and the modern number concept.[13] For the Arabs were the sole preservers of Greek culture for several centuries, and, when they eventually transmitted it to the West, it bore many distinctive marks of oriental thinking. The modern number concept as a thing in its own right must be compared with the synthesis the Arabs produced through amalgamation of Hellenism with Babylonian and Indian elements. In tracing the detail of this process of amalgamation, recent research has shifted more and more from the Syrian scene to that of eastern Persia.

As regards the number concept, however, this historical analysis, strange to say, is not needed. A glance into the *Almagest* of Ptolemy,[14] composed about A.D. 150, clearly shows that here, entirely within Greek culture, mathematicians used sexagesimal fractions with that same perfection characteristic of the modern number concept. Ptolemy used tables of sines, or, what is essentially the same, tables of chords for every $\frac{1}{2}°$ from $0°$ to $180°$, with chords in sexagesimal parts of the radius, for example, $\sin 1° = 1'2''50'''$. A third column listed the differences for interpolation, which shows that this, too, was in full use.

By the way, it is striking how superior the Greek system of numerals was to that of the Romans, that totally unmathematical people, whose preposterous numerals maintained themselves down to our day merely through the political power of the Imperium. The Greeks used the letters of the alphabet for number symbols: a, β, γ, . . . , ϑ for 1, 2, 3, . . . , 9; then ι, χ, λ, . . . , for 10, 20, 30; ρ, σ, τ, . . . , for 100, 200, 300, . . . ; 1,000 was designated again by the letter a, but with a little dash to the left and below ($'a$); similarly, $'\beta$ for 2,000; etc. These symbols differ in principle not so very much from our positional system. Only the zero was missing, and with it the advantage of purely positional computation, implying all the arithmetical schematization which we learn as children. It is very peculiar how Ptolemy used the symbol of o in the first column in the meaning it had for the Greeks, as 70°, and in the second and third columns, however, as that which we call 0 (zero); thus $0'3''15'''$ is completely positional writing. Since he never showed an actual numerical computation, we cannot see to what extent he used positional principles in computing.

From where did Ptolemy take the sexagesimal system of which he spoke as of something well known and established? This question becomes especially interesting if we examine how Archimedes handled similar problems. In measuring the circle—a problem with which the next section will deal, because of the infinite process which it involves—Archimedes struggles with complicated fractions, where Ptolemy uses sexagesimal, as we do decimal, fractions. Finally, he obtains the result

$$3\tfrac{10}{71} < \pi < 3\tfrac{10}{70},$$

which we should write in the form

$$3.1408 . . . < \pi < 3.1428 . . . .$$

With Hipparchus, whose sine tables are lost, but from whom Ptolemy doubtless took all that was best,[15] all astronomical data are still given as rational fractions reduced to lowest terms.

The question of the origin of the modern number concept has thus been well delimited as to when Greek mathematics incorporated this concept into its own system. With Ptolemy, at any rate, this incorporation has been achieved. Its Babylonian origin seems certain, provided we can really identify sexagesimal computations recorded on the clay tablets of Hammurabi. But it is only the fusion of sexagesimal computation with the powerful apparatus of Greek abstract mathematics, accomplished in the *Almagest*, which permitted the computation of tables of chords to two sexagesimals, which is equivalent to three decimals.

So much on the origin of the modern number concept and on the stages of its development. To compare it with the Greek number concept,[16] we must avoid two misunderstandings. First, that the Greeks ever designated as "number" anything of the nature we call "number." In Greek, ἀριθμός (*arithmos*) means a "whole number." Calculation with fractions is clothed in the language of proportions. The other misunderstanding is the assumption that line segments, areas, time intervals—in short, all that the Greeks included under the generic term "quantities" (μεγέθη)—were a correlative to our number concept, in the sense in which we assign to such entities numerical measures—measure them in infinite decimals. The Greek counterparts of the infinite decimals are rather the λόγοι (*logoi*), the "ratios," which are the objects of the theory of proportions described above and, in fact, of all the infinite processes mentioned so far. If we speak today of the numerical measure of an area, we mean in fact the number which indicates the ratio between that area and a given unit square, that is, the ratio between two areas. The Greek definition of ratio given earlier, however, was so skilful that the ratio of two line segments, on the one hand, could be compared to the ratio of two areas, on the other hand, as we saw in the case of triangles of equal altitudes whose areas had the same ratio as their bases. In that way the Greek concept of "ratio" transcended the sphere of special applications. There was no need for separate theories of segments, of areas of plane figures, etc. With the general theory of quantities the ascent to an abstract system of mathematics had been achieved. The objects of this abstract mathematics, the λόγοι, by means of which the Greeks had found and expressed many important mathematical relations, play the role of the modern number concept. The principal difference is that, until the indicated development with Ptolemy, they did not add and multiply these λόγοι. Still, they performed analogous operations which fully enabled them to make all important discoveries, but they lacked the flexible forms that render the modern number concept so convenient a tool even in the hands of the non-mathematician. Indeed, it seems as though the Greeks had been searching for an access to this abstract mathematics—the "arithmetization" of mathematics—through a new number concept, at the very time Plato intervened in the epistemological development of mathematics, perhaps with an even stronger hand than can be precisely proved today. And it seems that here too—as in cosmology—Aristotle called a disastrous halt. That halt may have been justified at the then particular state of scientific inquiry, but owing to Aristotle's towering authority, it remained effective down to distant later times when the state of scientific development imperatively demanded new approaches.

5. ARCHIMEDES' MEASUREMENT OF THE CIRCLE AND THE SINE TABLES

The measurement of the circle by Archimedes,[17] which we just touched upon in a different connection, contains one of the most characteristic instances of an infinite process, and it occupies a unique position among Archimedes' works. While most of his theses treat of quadratures, cubatures, and centroid determinations, this work has nothing whatever to do with squaring of the circle in the sense of Hippocrates, although it is sometimes mistakenly described as such. For the question is not how to construct with compass and ruler a square exactly equal to a circle but how to find fractions with lowest possible numerators and denominators which should equal the ratio between the circle and the square over its radius as closely as possible. We are not so much interested in the result Archimedes attained—$3\frac{10}{71} < \pi < 3\frac{10}{70}$—and which Apollonius is said to have further improved but rather in the infinite process which Archimedes discovered for computing π as accurately as we want.

Archimedes started with the idea suggested earlier by Antiphon (see our discussion in Sec. 3). He considered the polygons of 6 sides, 12 sides, 24 sides, etc., inscribed in and circumscribed about the unit circle, the perimeters of the former all being less than 2π, those of the latter greater than 2π.* His new discovery is the rule for producing from the perimeters of the inscribed and circumscribed n-sided polygons those of the $2n$-sided polygons by using only compass

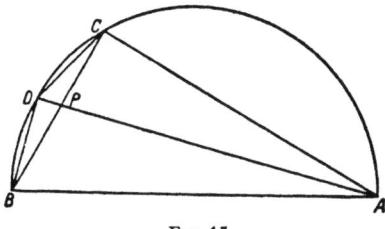

FIG. 15

and ruler (or, as we would say, by using only the four elementary operations and the extraction of square roots). Since the usual accounts of his reasoning, employing modern terminology, fail to bring out clearly the point in which we are interested, we shall briefly explain what Archimedes really did.

Let AB in Figure 15 be a diameter of the circle, $BC = s_n$ the side of the inscribed n-gon, and $BD = s_{2n}$ the side of the inscribed $2n$-gon. Hence AD is the bisector of angle BAC. It then follows from elementary geometry that the right triangles ABD, BPD, and APC are all similar to one another. From the similitude of the first two follows

$$\frac{AB}{AD} = \frac{BP}{BD};$$

from the similitude of the first and third,

$$\frac{AC}{AD} = \frac{PC}{BD}.$$

Hence, by addition,

$$\frac{AB + AC}{AD} = \frac{BP + PC}{BD};$$

* [Toeplitz has here shifted from considering areas to perimeters.—Editor.]

or, since $BP + PC = BC$,

$$AD:BD = (AB + AC):BC$$

for the squares; therefore,

$$AD^2:BD^2 = (AB + AC)^2:BC^2 \; ;$$

and by corresponding addition

$$(AD^2 + BD^2):BD^2 = (AB^2 + AC^2 + 2AB \cdot AC + BC^2):BC^2 \; ;$$

or, from the theorem of Pythagoras,

$$AB^2:BD^2 = (2AB^2 + 2AB \cdot AC):BC^2 \; .$$

If the radius of this circle is unity, and hence $AB = 2$, this becomes

$$4:s_{2n}^2 = (8 + 4AC):s_n^2 \; ;$$

or, since $AC^2 = AB^2 - BC^2 = 4 - s_n^2$,

$$s_{2n}^2 = \frac{s_n^2}{2 + \sqrt{4 - s_n^2}} . \tag{5.1}$$

This is the formula which Archimedes used to obtain the side of the $2n$-gon from that of the n-gon. He began with $s_6 = 1$ and obtained from it

$$s_{12}^2 = \frac{1}{2 + \sqrt{3}} = 2 - \sqrt{3}, \qquad \text{or} \qquad s_{12} = \sqrt{2 - \sqrt{3}} .$$

Then he used $\sqrt{3} < 1{,}351/780$ (how he arrived at this approximation he did not say) and obtained from that a lower bound for s_{12}, hence for $12s_{12} = u_{12}$, the perimeter of the inscribed 12-gon, and, since the circumference of the circle is greater, a lower bound also for the circumference of the circle. We see how much effort and ingenuity he had to use in calculating with fractions, where we with our decimals and Ptolemy with his sexagesimals can proceed schematically. The important thing, however, is that his procedure can be indefinitely continued. From s_{12} he can with the same operations derive s_{24} and likewise s_{48} from s_{24}, and thus we can continue to any desired degree of approximation.

Analogously, Archimedes derived through elementary reasoning for the side t_{2n} to the circumscribed $2n$-gon a formula in terms of t_n,

$$\frac{2}{t_{2n}} = \frac{2}{t_n} + \sqrt{1 + \left(\frac{2}{t_n}\right)^2} , \tag{5.2}$$

and obtains from this an upper bound for 2π, which together with previous lower bound, continued to the 96-gon, gives him the result stated above.

Here, for the first time, a principle is given by means of which a quantity, which cannot be computed exactly, could be approximated to an arbitrary accuracy; or, as we would say, "to any number of decimals," or, as the Greeks would say, "with an error less than any arbitrarily small preassigned ϵ."

So far the measurement of the circle appears as a mere isolated problem, of interest chiefly because leading to the computation of π. In fact, it was far more than that. In modern symbols we have

$$s_n = 2 \sin \frac{\pi}{n}, \qquad t_n = 2 \tan \frac{\pi}{n}.$$

Formula (5.1) therefore gives sin $a/2$ in terms of sin a, formula (5.2) gives cot $a/2$ in terms of cot a. But (5.1) does not have the exact form we use when, by rationalizing the denominator, we write

$$s_{2n}^2 = 2 - \sqrt{4 - s_n^2} ; \tag{5.1a}$$

also, we often express sin a in terms of sin $a/2$,

$$\sin a = 2 \sin \frac{a}{2} \sqrt{1 - \sin^2 \frac{a}{2}},$$

or

$$s_n = s_{2n} \sqrt{4 - s_{2n}^2}, \tag{5.1b}$$

from which (5.1a) follows at once by solving for s_{2n}^2. Thus what Archimedes was developing here was the beginnings of trigonometry. From recently discovered Arabic translations of Archimedes[18] we know that he went much further—that he knew not only the form (5.1a) of formula (5.1) but also a general theorem closely related to the addition theorem for sin $(a + \beta)$ and sin $(a - \beta)$.

With Ptolemy we find the addition theorem itself in the form in which we still express it in school mathematics as "Ptolemy's Theorem" concerning a quadrilateral inscribable in a circle. We find it there, however, not as a theorem in its own right but in the service of the task which astronomy was setting for the mathematicians of the next fifteen hundred years—to compute sine tables. The true significance of Archimedes' measurement of the circle is that it furnished the starting point for this computation. As described by Ptolemy, a sine table is computed as follows: sin 36° is known from the construction of the regular pentagon; by virtue of Archimedes' formula (5.1a), sin 18° may be computed; next sin 30° is given, and hence sin 15°. According to the addition theorem, sin 3°, sin 1½°, and sin ¾° may be obtained. The second auxiliary is the formula

$$\frac{\sin x}{\sin y} < \frac{x}{y},$$

when $0° < y < x < 90°$. This formula had been used already by Aristarchus, the first proponent of the heliocentric system, and Archimedes cited it as "Aristarchus' Formula":

$$\frac{\sin 1°}{\sin \frac{3}{4}°} < \frac{1}{\frac{3}{4}} = \frac{4}{3},$$

$$\frac{\sin 1\tfrac{1}{2}°}{\sin 1°} < \frac{1\tfrac{1}{2}}{1} = \tfrac{3}{2} \, ;$$

hence

$$\tfrac{2}{3} \sin 1\tfrac{1}{2}° < \sin 1° < \tfrac{4}{3} \sin \tfrac{3}{4}° \, .$$

And, since Ptolemy had already computed $\sin 1\tfrac{1}{2}°$ and $\sin \tfrac{3}{4}°$ in sexagesimals, he formulated

$$1'2''50''' < \sin 1° < 1'2''50\tfrac{2}{3}''' \, ,$$

or, as written in Ptolemy's characteristic manner,

$$1'2''50''' < \sin 1° < 1'2''50''' \, . \tag{5.3}$$

From this he obtained $\sin \tfrac{1}{2}°$, and next, with the aid of the addition theorem, the sines of the multiples of $\tfrac{1}{2}°$. With the sine of each angle his table indicated also the thirtieth part of the difference from the sine of the angle increased by $\tfrac{1}{2}°$ to compute interpolations.

Many of these methods of computation, this first tabulation of a function, were no doubt contained already in Archimedes' *Book of Circles*, of which the extant "Measurement of the Circle" may be only a part, and also in the twelve books of chord tables of Hipparchus (about 150 B.C.). Ptolemy contributed perhaps only the sexagesimal notation and the greater accuracy which it made possible. One thing he did contribute, at any rate, which was a novelty in Greek mathematics—the manner of writing formula (5.3), in which he simply omitted all higher sexagesimals, writing an inequality which, taken literally, is absurd and means only an approximation. The contrast to the strictly exact inequalities of Archimedes is most striking.

With Ptolemy there then emerged that variety of the modern number concept which guided for centuries the minds of later mathematicians and still dominates modern "applied" mathematics—the idea of number as an approximation. In the same sense Briggs, around 1620, recalculated for the base of 10 Napier's logarithm tables, which were still strictly computed, neat inequalities. Briggs, too, started discarding the higher places and so, while shortening the calculations, deprived the tables of some of their accuracy. Well aware of that, he justified himself by the fact that this *morbus decimalium*, this "disease of the decimals," had been present in the sine tables long before him. The origin of this usage we found in Ptolemy.

6. THE INFINITE GEOMETRIC SERIES

The infinite decimal fraction has already occupied our attention. We encountered it as a formal representation originating in the measurement of a quantity in terms of a unit quantity of the same kind. Stevin was aware that for a quantity equal to $\tfrac{5}{6}$ of unity, for example, the decimal does not terminate, but becomes .83333 . . . , and we have decided to accept this formal representation as the substratum of the idea of number, for we know how to add, subtract, multiply, and divide two infinite decimal fractions.

We may, however, regard an infinite decimal fraction from another point of view. Just as 0.83 is nothing else but $8/10 + 3/10^2$, we may consider also the sum of infinitely many terms

$$\frac{8}{10} + \frac{3}{10^2} + \frac{3}{10^3} + \cdots$$

and examine whether this infinite process has a meaning and, if so, what it is.

For this we need a lemma, with which the Greeks already were quite familiar, namely, *the theorem of the geometric series:*

$$\frac{a^{n+1} - b^{n+1}}{a - b} = a^n + a^{n-1}b + \ldots + a b^{n-1} + b^n. \tag{6.1}$$

The proof is very simple. The product of the denominator and the right member simply is

$$a^n(a - b) + a^{n-1}b(a - b) + \ldots + b^n(a - b)$$

$$= a^{n+1} - a^n b + a^n b - a^{n-1}b^2 + \ldots + a b^n - b^{n+1} = a^{n+1} - b^{n+1},$$

which is the numerator. For $a = 1$, $b = x$, $n = m - 1$, we obtain the formula which we learn in school:

$$\frac{1 - x^m}{1 - x} = 1 + x + \ldots + x^{m-1}. \tag{6.2}$$

Now we need a second lemma, which is a consequence of the first, namely:

If $0 \leq a < 1$, then a^n approaches zero as n increases; that is, if an arbitrarily small positive number ϵ is preassigned, n can be chosen so that $a^n \leq \epsilon$.

For $a \leq \frac{1}{2}$ this is nothing else but the continuity axiom in the second formulation which we derived above from the original one. If a is closer to 1, that is, $a > \frac{1}{2}$, we can prove the lemma quite similarly with the theorem of the geometric series. For if $a < 1$, then $x = 1/a > 1$, and if we apply formula (6.2) to this x—leaving m so far arbitrary—each term on the right side is greater than or equal to 1, and hence the sum of all terms is greater than or equal to m; that is,

$$\frac{x^m - 1}{x - 1} \geq m, \quad \text{or} \quad x^m \geq 1 + m(x - 1).$$

So far m is arbitrary. However, according to the continuity axiom in its original formulation, m can be chosen so great that m times the given quantity $x - 1$ will exceed any preassigned number, such as the reciprocal $1/\epsilon$ of the preassigned arbitrarily small ϵ mentioned in the lemma above. Then, a fortiori, $x^m > 1/\epsilon$, and, consequently, $a^m = 1/x^m < \epsilon$; and it was the existence of such an exponent which we had to demonstrate.

Now we have assembled all the tools needed to proceed with the summation of infinitely many terms. By summing up m terms, we have

$$s_m = 1 + x + x^2 + \ldots + x^{m-1} = \frac{1}{1-x} - \frac{1}{1-x} x^m.$$

Hence the difference is

$$\frac{1}{1-x} - s_m = \frac{1}{1-x} x^m.$$

If $0 \leq x < 1$, then, according to the second lemma, m can be taken so large that not only x^m but also $[1/(1-x)]x^m$ becomes smaller than any preassigned ϵ. Hence, for the number m of terms sufficiently large, s_m differs from $1/(1-x)$ by as little as we please. The number $1/(1-x)$ is therefore the result toward which, with any required accuracy, the infinite process of summation is leading.

THEOREM OF THE INFINITE GEOMETRIC SERIES: *If $0 \leq$ x < 1, the sum $1 + x + x^2 + \ldots, + x^n$ comes ever closer to $1/(1 - $ x$)$ as* n *increases.*

For this we write briefly, and in a manner that cannot be misunderstood:

$$1 + x + x^2 + \ldots = \frac{1}{1-x}.$$

7. CONTINUOUS COMPOUND INTEREST

The Swiss mathematician Jakob Bernoulli, who lived around 1700, once proposed the following problem: "Quaeritur, si creditor aliquis pecuniam suam foenori exponat, ea lege, ut singulis momentis pars proportionalis usurae annuae sorti annumeretur; quantum ipsi finito anno debeatur?"[19] The creditor, of whom Bernoulli speaks, thus lends money at interest under the condition that during each individual moment the proportional part of the annual interest be added to the principal. Let us explain this more fully: Our savings banks compound, at the end of the year, the interest earned. If at the beginning of the year, the depositer puts a dollars into a savings account, and if the rate of interest is 5 per cent, the bank adds at the end of the year 5 per cent of a, or $(1/20)a$, to the principal, so that the depositor then has

$$a + \tfrac{1}{20} a = a \left(1 + \tfrac{1}{20}\right)$$

dollars in his account. Other institutions act differently. Some discount banks add the interest semiannually. Under this plan a sum a deposited at 5 per cent interest will after six months have grown to

$$b = a + \tfrac{1}{2} \cdot \tfrac{1}{20} a = a \left(1 + \tfrac{1}{2} \cdot \tfrac{1}{20}\right),$$

and, if during the second half year this increased amount earns interest, the balance at the end of the full year is

$$b \left(1 + \tfrac{1}{2} \cdot \tfrac{1}{20}\right) = a \left(1 + \tfrac{1}{2} \cdot \tfrac{1}{20}\right)^2.$$

Similarly, if a bank adds the interest due every quarter, the balance at the end of the year is

$$a \left(1 + \tfrac{1}{4} \cdot \tfrac{1}{20}\right)^4 ;$$

if interest is added every month, the balance is

$$a \left(1 + \tfrac{1}{12} \cdot \tfrac{1}{20}\right)^{12} ;$$

and so on. It is clear that the arrangement is the more advantageous for the investor the more frequently interest is added. If this is done n times a year, and if the rate of interest is not $5/100$ but more generally x, the sum of a dollars deposited at the beginning of the year will be worth

$$a \left(1 + \frac{x}{n}\right)^n$$

at the end of the year.

The question Bernoulli proposed amounted to this: Whether through contracts specifying ever larger n the depositor could acquire fortunes no matter how large; or, to put it mathematically, whether for a fixed x the value of

$$s_n = \left(1 + \frac{x}{n}\right)^n$$

would increase without limit as n increases without limit.

To investigate this question, we begin by choosing $x = 1$, which would mean an interest rate of 100 per cent. This, of course, is not a very realistic assumption, but it simplifies the calculation, and subsequently we can readily adapt our results to other values of x. By diligent computation we find

$$s_1 = 2.000 ; \quad s_2 = 2.250 ; \quad s_3 = 2.370 ; \quad s_4 = 2.441 ; \quad s_5 = 2.488 ;$$

$$s_6 = 2.522 ; \quad \ldots ; \quad s_{10} = 2.594 ; \quad \ldots ; \quad s_{100} = 2.704 ; \quad \ldots .$$

These numbers show the increase of s_n with n, as was to be expected. But the increase is gradual only. Although we went up to $n = 100$, we still are well below 3.000. Doubtless no computation, no matter how tireless, will give us a decisive answer to the question of whether s_n will ever grow beyond 3.000, and beyond any limit, or whether it will always stay below 3.000 no matter how large we take n. The answer cannot be obtained through computations no matter how assiduous but only through mathematical reasoning.

The idea which leads to the desired goal, we grant, is not exactly obvious. We begin by examining our expression s_n for negative values of n. Let

$$t_n = \left(1 - \frac{1}{n}\right)^{-n} ;$$

t_1 then is meaningless, but

$$t_2 = (1 - \tfrac{1}{2})^{-2} = (\tfrac{1}{2})^{-2} = 2^2 = 4 .$$

Similarly, we find

$$t_3 = 3.375 ; \quad t_4 = 3.161 ; \quad t_5 = 3.052 ; \quad t_6 = 2.986 ; \quad \ldots ;$$

$$t_{10} = 2.868 ; \quad \ldots ; \quad t_{100} = 2.732 ; \quad \ldots .$$

As in the case of the s_n above, these t_n are given only to three decimals, discarding the higher places.

These numerical results suggest that perhaps t_n decreases with increasing n, just as the s_n were increasing; also that each t_n is greater than the corresponding s_n. If we could *prove* this observation, which so far is merely empirical, we would have shown that Bernoulli's moneylender would not acquire unlimited fortunes. Indeed, it would then be clear that all s_n beyond s_{100}, as well as all t_n beyond t_{100}, would begin with 2.7 and would differ only in the higher decimals.

But how do we prove this? Very simply: We have

$$1 + \frac{1}{n} = \frac{n+1}{n} = \frac{n+1}{(n+1)-1} = \left[\frac{(n+1)-1}{n+1} \right]^{-1} = \left(1 - \frac{1}{n+1} \right)^{-1} .$$

Hence

$$\left(1 + \frac{1}{n} \right)^{n+1} = \left(1 - \frac{1}{n+1} \right)^{-(n+1)} ;$$

that is,

$$s_n \left(1 + \frac{1}{n} \right) = t_{n+1}, \quad \text{or} \quad s_n + \frac{1}{n} \, s_n = t_{n+1} .$$

This shows that for every n

1. $$s_n < t_{n+1} .$$

The first part of our conjecture has thus been proved. But this does not yet prove that s_n could not grow without limit. This too will soon be established.

For every n,

2. $$s_n < s_{n+1} .$$

This we can derive from the theorem of the geometric series of Section 6 as follows: First, for the special assumption $0 < b < a$, we obtain from that theorem

$$(n+1) b^n < \frac{a^{n+1} - b^{n+1}}{a - b} < (n+1) a^n ; \tag{7.1}$$

for, on the one hand, the $n+1$ terms of the "geometric series" are, except the last one, all greater than b^n; on the other hand, all except the first one are smaller than a^n. If we place next

$$a = 1 + \frac{1}{n}, \quad b = 1 + \frac{1}{n+1},$$

whereby $0 < b < a$, we obtain from inequality (7.1)

$$\frac{[1 + (1/n)]^{n+1} - [1 + (1/n+1)]^{n+1}}{[1 + (1/n)] - [1 + (1/n+1)]} < (n+1) \left(1 + \frac{1}{n} \right)^n .$$

Simplifying the denominator on the left side to

$$\frac{1}{n} - \frac{1}{n+1} = \frac{1}{n(n+1)},$$

and multiplying the inequality by it, we obtain

$$\left(1 + \frac{1}{n}\right)^{n+1} - \left(1 + \frac{1}{n+1}\right)^{n+1} < \frac{1}{n}\left(1 + \frac{1}{n}\right)^n;$$

that is,

$$s_n\left(1 + \frac{1}{n}\right) - s_{n+1} < \frac{1}{n} s_n, \qquad \text{or} \qquad s_n < s_{n+1}.$$

Quite similarly we can prove

3. $$t_n > t_{n+1}.$$

This time we place

$$a = 1 - \frac{1}{n+1}, \qquad b = 1 - \frac{1}{n},$$

wherefore again $0 < b < a$. As before, we obtain from inequality (7.1), this time from its left side,

$$\frac{1}{n}\left(1 - \frac{1}{n}\right)^n < \left(1 - \frac{1}{n+1}\right)^{n+1} - \left(1 - \frac{1}{n}\right)^{n+1},$$

which leads at once to $t_n^{-1} < t_{n+1}^{-1}$, which is equivalent to 3.

The rest of our proof follows from these three facts by reasoning rather than by calculation.

4. $$\text{Every } s_a < \text{every } t_b .$$

For if $b = a + 1$, then, according to 1, $s_a < t_{a+1} = t_b$. Suppose $b > a + 1$; according to 1, $s_a < t_{a+1}$; since, according to 3, $t_b > t_{a+1}$, this means $s_a < s_{b-1}$, and this, according to 1, means $s_a < t_b$. The statement is thus true for all a and b.

Finally, we prove the following:

5. If ϵ is any preassigned positive number, then there exists a number p such that, for $n \geq p$,

$$t_{n+1} - s_n \leq \epsilon.$$

First, in deriving 1, we saw that

$$s_n + \frac{1}{n} s_n = t_{n+1}.$$

Hence we can write in our inequality to be proved

$$t_{n+1} - s_n = \frac{1}{n} s_n.$$

Moreover, since, according to **4**, every $s_n < t_2 = 4$, we obtain

$$t_{n+1} - s_n \leq \frac{4}{n}.$$

For this to be less than ϵ, one need only choose n so large that $4/n < \epsilon$, that is, to take $p > 4/\epsilon$.

With all these facts concerning continuous compound interest established, we now recognize the similarity with the measurement of the circle. If the side of the hexagon inscribed in the circle is s_1, that of the 12-gon s_2, that of the 24-gon s_3, etc., and that of the circumscribed hexagon t_1, that of the circumscribed 12-gon t_2, etc., it is evident that the s_n grow ever larger and the t_n ever smaller; that any s_a is smaller than any t_b; and that the s_n and t_n approach each other closer and closer as n increases; that they agree in more and more decimal places, or, differently expressed, that for any preassigned small ϵ we can find a p so large that for $n \geq p$ the difference $t_n - s_n \leq \epsilon$. All these facts we found to exist in the case of continuous compound interest too, but here we established them by calculations rather than through observation. If we assign to ϵ successively values $\frac{1}{10}$, $\frac{1}{100}$, ..., then for $\epsilon = \frac{1}{100}$, for example, from a definite index on all s_a will differ from each other by less than $\frac{1}{100}$, and so will all t_b, and likewise all s_a will differ from all t_b by less than $\frac{1}{100}$. They will agree in the first two decimal places; and, similarly, from a certain larger index on they will agree in the first three decimals and so on. (A possible exception to this we shall mention in the next section.) In this manner a certain infinite decimal fraction "consolidates" itself, and it takes nothing but a little patience to compute it to several decimals. Doing so, we find it beginning with 2.718 . . . .* Likewise Archimedes showed how to compute the area of the circle to as many decimals as is desired and that it was merely a question of patience actually to obtain eight or thirteen or any desired number whatever of decimal places. It interests us here to demonstrate what is common to both problems.

8. PERIODIC DECIMAL FRACTIONS

In this whole course we take up topics which we encountered in elementary-school mathematics and treat them from a new point of view. There we often find ourselves in the position of a violin student heretofore taught superficially for mere amateur playing; for serious, professional playing he has to start over again from the beginning. So we have "to start at the very beginning."

What does 3.907 mean? Very simply, it means

$$3 + \frac{9}{10} + \frac{0}{100} + \frac{7}{1,000}, \quad \text{or} \quad \frac{3,907}{1,000};$$

that is, a fraction whose denominator is a power of ten. This seems to say that only very special fractions are at the same time decimal fractions. However, the

---

* [This number, which is both $\lim\limits_{n \to \infty} [1 + (1/n)]^n$ and $\lim\limits_{n \to \infty} [1 - (1/n)]^{-n}$, is designated by the symbol e.—EDITOR.]

question whether all fractions can be written in the form of decimal fractions is by no means that simple. For example, $\frac{2}{5}$ is a fraction whose denominator is not a power of ten, but, by multiplying numerator and denominator by 2, it becomes $\frac{4}{10}$, or 0.4, a decimal fraction.

It is apparent that the same principle can be applied whenever the denominator has only factors 2 and 5. For example,

$$\frac{3}{20} = \frac{3}{2 \cdot 2 \cdot 5} = \frac{3 \cdot 5}{2 \cdot 2 \cdot 5 \cdot 5} = \frac{15}{10 \cdot 10} = 0.15 .$$

Generally,

$$\frac{a}{2^p \cdot 5^q} = \frac{a \cdot 2^{q-p}}{2^p \cdot 5^q \cdot 2^{q-p}} = \frac{a \cdot 2^{p-q}}{2^q \cdot 5^q} = \frac{a \cdot 2^{p-q}}{10^q} , \qquad \text{for } q > p ;$$

$$\frac{a}{2^p \cdot 5^q} = \frac{a \cdot 5^{p-q}}{2^p \cdot 5^q \cdot 5^{p-q}} = \frac{a \cdot 5^{p-q}}{2^p \cdot 5^p} = \frac{a \cdot 5^{p-q}}{10^p} , \qquad \text{for } p > q .$$

In case $p = q$ the denominator of the given fraction can at once be written as a power of ten. Thus *any fraction, and only such, whose denominator contains no factors other than 2's and 5's can be written as a decimal fraction*, for, by multiplying numerator and denominator, the latter becomes a power of ten. If there is some other factor besides 2's and 5's in the denominator, say, a factor 3, it can never be made to disappear, and hence no power of ten can result by multiplication. We have thus come to know the set of fractions which can be written as decimal fractions. In all this we have had in mind only "finite decimal fractions" which "terminate"; no "infinite decimal fractions" have as yet been brought into the game. We shall approach these but cautiously, for we do not even know what they are until we realize that any infinite decimal fraction means an infinite process.

We still need a preliminary to bring out this infinite process with full clarity. The method used above to transform a fraction like $\frac{3}{20}$ into a decimal can be replaced by one which has the advantage of being applicable also to fractions which cannot be transformed into finite decimals. This is the method of division. This process, the mechanics of which we learn as children in school, is used at first when the denominator is a divisor of the numerator, as for $\frac{460}{20} = 23$.

(Applying the same process to $\frac{3}{20}$, we find that 20 divides into 3 zero times; into 30 it divides one time with a remainder 10; into 100 it divides five times without a remainder. The same result, $\frac{3}{20} = 0.15$, has thus been obtained, but in a very different manner which is computationally perhaps more convenient.

Now nothing can keep us from applying this division method to a fraction like $\frac{4}{3}$. We begin with "3 divides into 4 one time." This says that $\frac{4}{3}$ lies between the two whole numbers 1 and 2, so that a decimal expansion—if there is one— must begin with 1 and nothing else. The second step will find the next decimal place. The second decimal must show how many full tenths are in the difference between $\frac{4}{3}$ and 1, that is, in $\frac{4}{3} - 1$. Instead we may say as well: by multiplying this difference by 10, how many units—not tenths—are contained in it?) Clearly,

$$3 < 10 \left(\tfrac{4}{3} - 1\right) = \tfrac{10}{3} < 4 .$$

Hence the second place is 3; that is, the decimal fraction which we are trying to find—if it exists—would begin with 1.3. But we know that it cannot exist! Strictly speaking, we have only shown that $1.3 < \tfrac{4}{3} < 1.4$, and that is exact.

In determining the next place, we encounter something that is new in principle. We have to ascertain how many hundredths are contained in the difference $\tfrac{4}{3} - 1.3$, or, also, how many units in one hundred times this difference. Now, obviously,

$$3 < 100 \cdot \left(\tfrac{4}{3} - 1.3\right) = 100 \cdot \frac{4 - 3.9}{3} = \tfrac{10}{3} < 4 .$$

This means that the next decimal place is again 3—that $1.33 < \tfrac{4}{3} < 1.34$. But more than that! One hundred times the difference between $\tfrac{4}{3}$ and 1.3 had exactly the same value as ten times the difference between $\tfrac{4}{3}$ and 1, namely, $\tfrac{10}{3}$. The remainder of 40 divided by 3, or, what is the same, $40 - 39$, is equal to 1, just as is the remainder of 4 divided by 3, which is $4 - 3$. Similarly, the remainder of 400 divided by 3 will again be 1, since $400 - 399 = 400 - 390 - (10 - 1)$, and so on. In other words, the elementary process of long division shows at once that all further places must be 3. Hence

$$1 < 1.3 < 1.33 < 1.333 < \ldots < \tfrac{4}{3} < \ldots < 1.334 < 1.34 < 1.4 < 2 .$$

This has taught us two things: (1) that the division process we learned in school, applied to $\tfrac{4}{3}$, leads to a decimal fraction all of whose decimal places are 3's and which has no end and (2) that $\tfrac{4}{3}$ lies between two terminating decimals as shown above. We must next generalize the results found in this special instance.

We begin by taking, instead of $1.333 \ldots$, any infinite decimal fraction, for example, $0.7424242 \ldots$, in which, aside from the first place, the group of digits 42 is repeated without end. Such an infinite decimal has, so far, no meaning. By using our previous results, however, we may make the following statements:

$$0.742 = 0.7 + \frac{42}{10^3},$$

$$0.74242 = 0.7 + \frac{42}{10^3} + \frac{42}{10^5} = 0.7 + 42 \left(\frac{1}{10^3} + \frac{1}{10^5}\right), \text{ etc.}$$

Continuing in this way, we obtain ever longer stretches of the geometric series

$$\frac{1}{10^3} \left(1 + \frac{1}{10^2} + \cdots\right),$$

and these, as we already know, approach ever closer to the value

$$\frac{1}{10^3} \cdot \frac{1}{1 - (1/10^2)} = \frac{1}{10} \cdot \frac{1}{10^2 - 1} = \frac{1}{990}.$$

Hence the values 0.742, 0.74242, . . . , come ever closer to the value

$$0.7 + 42 \cdot \tfrac{1}{990},$$

that is, to a definite rational number, a fraction. And, since our reasoning was in no way restricted to any particular infinitely repeating decimal, it would be equally valid for any other, say, 0.46932932932 . . . . We have thus established this result:

THEOREM I. *If in an infinite "periodic" decimal fraction we consider its "partial fractions," that is, the finite decimal fractions which are obtained by terminating it somewhere, their values come ever closer to a definite value, which is a rational number.*

The next question is whether the reverse of this statement is also true; that is, whether any rational number, if transformed into a decimal, is always a periodic decimal. To show this next, we must remind ourselves for a moment, in a little detail, of the rule for long division which we learned as children. How, for example, do we apply this rule of division to $\tfrac{3}{14}$? Fourteen divides into 3 zero times, remainder 3; 14 divides into 30 two times, remainder 2; 14 divides into 20 one time, remainder 6; and so on. Now we must clearly realize one thing; all that follows in continuing with the division from here on depends only on this last remainder 6. We could continue by merely keeping in mind the last remainder 6, forgetting all previously obtained quotients and remainders. For, in continuing, we have: 14 goes into 60 four times, remainder 4; 14 goes into 40 two times, remainder 12; etc. Again the whole continuation from here on is determined only by the last remainder 12; 14 goes into 120 eight times, remainder 8; etc. Continuing yet further, we find: 14 goes into 80 five times, remainder 10; 14 goes into 100 seven times, remainder 2. At this moment we have obtained a remainder 2 which we had before, namely, after the second division. Since, however, the whole further continuation depends only on this last remainder 2, all that follows must be a repetition of what followed the second step. Writing the quotients and remainders in the order obtained, the quotients in the customary way, and the remainders, respectively, below them

$$0. \quad 2 \mid 1 \ 4 \ 2 \ 8 \ 5 \ 7 \mid 1 \ 4 \ 2 \ldots$$

$$3 \mid 2 \quad 6 \ 4 \ 12 \ 8 \ 10 \mid 2 \quad 6 \ 4 \ 12 \ldots,$$

we see how the decimal fraction becomes indeed "periodic" at this place, that is, the six quotients 1, 4, 2, 8, 5, and 7 repeat forever.

The same reasoning evidently applies to every fraction, as it did to $\tfrac{3}{14}$. In the case of any fraction p/q there are, instead of the fourteen remainders 0, 1, 2, 3, . . . , 13 and q remainders 0, 1, 2, 3, . . . , $q - 1$. The remainders which actually occur in dividing p by q are some of these q possibilities. In the case of $\frac{3}{14}$ six of the possible fourteen actually occurred. Exactly how many there will occur in the case of p/q cannot be said in general. Since, however, only q different remainders are available, the process of division continuing forever, after $q + 1$ divisions at least two equal remainders must have turned up. As soon, however, as a remainder equals a former one, the process becomes periodic, beginning immediately after the place where the first of the two equal remainders appeared. This is what we had set out to show:

THEOREM II. *The division rule applied to any fraction leads to a periodic decimal fraction.*

Now let us check our two results against each other. The rule for long division applied to p/q yields a periodic decimal fraction. Such periodic decimal can be infinite, or terminating like 0.8. But 0.8 is in fact the same as 0.8000000000 . . . and thus periodic with the repeating one-digit period 0. According to Theorem I, this periodic decimal has a "value" which is rational; that is, its partial fractions approach a definite rational number. Now let us be very conscientious and ask: Is the "value" of the decimal fraction the same as the p/q from which we started the division? In the example, is the "value" of 0.2142857142857 . . . really $\frac{3}{14}$? Most students might be inclined to regard this as obvious, and the question as superfluous. It is, in fact, not difficult to prove. But that it is by no means self-evident appears from the following observation. What is the value of the infinite decimal fraction 0.7999999 . . . ? It is

$$0.7 + \frac{9}{10^2} + \frac{9}{10^3} + \ldots = 0.7 + \frac{9}{10^2}\left(1 + \frac{1}{10} + \frac{1}{10^2} + \ldots\right)$$

$$= 0.7 + \frac{9}{10^2} \cdot \frac{1}{1 - (1/10)} = 0.7 + \frac{9}{10^2} \cdot \frac{10}{9}$$

$$= 0.7 + 0.1 = 0.8 = 0.8000 \ldots .$$

That is, its "value" is $\frac{4}{5}$. If, however, we apply long division to $\frac{4}{5}$, we obtain not 0.79999 . . . but 0.8. Thus we have here two distinct decimal fractions with the same "value." That shows that we may not simply interchange a decimal fraction and its "value."

We can easily see, however, that this difficulty is not serious. It happens only in decimal fractions in which the period consists of 9s only. Otherwise two different decimal fractions cannot have the same "value," and two different "values" cannot belong to one and the same decimal fraction.

The results of this discussion incidentally show in an altogether new way that there are irrational numbers. Consider the following infinite decimal fraction: 0.10100100010000 . . . , in which the 1's are separated successively by one, two, three, four, . . . , 0's. It is not periodic; hence it cannot represent a rational number. Its value* must therefore be irrational.

* [It has not been shown to have one.—EDITOR.]

9. CONVERGENCE AND LIMIT

Let us now abstract the essence of all that we have come to know in the variety of examples discussed above and formulate the situations and facts which we encountered again and again as general concepts and theorems.

DEFINITION OF LIMIT. *If there is given a sequence of infinitely many numbers* s_1, s_2, . . . , *and if for any preassigned number* $\epsilon > 0$ *a number* s_p *of the sequence can be found such that it and all subsequent* s_n, $(n \geq p)$, *differ from a fixed number* s *by less than* ϵ, *then we say that the sequence converges to the limit* s, *and write*

$$\lim_{n\to\infty} s_n = s .$$

(The abbreviation "lim" is from the Latin word *limes*, meaning boundary or limit.)

We see that this is the concept to which we were led again and again in discussing our various examples. But these examples did more than familiarize us with this concept. We shall recognize repeated experiences of ours also in the following two theorems:

THEOREM I. *Let there be two sequences* s_1, s_2, . . . , *and* t_1, t_2, . . . , *which satisfy these four conditions: (1)* $s_\alpha \leq s_{\alpha+1}$; *(2)* $t_\beta \geq t_{\beta+1}$; *(3)* $s_\alpha \leq t_\beta$; *(4) for any preassigned* $\epsilon > 0$ *there is a* p *such that, for all* $n \geq p$, $t_n - s_n \leq \epsilon$. *Then both sequences are convergent, and both converge to the same limit:*

$$\lim_{n\to\infty} s_n = \lim_{n\to\infty} t_n .$$

To formulate the second theorem, we first define two new terms: (1) If in a sequence s_1, s_2, . . . , the condition holds that $s_n \leq s_{n+1}$, the sequence is called *monotonic increasing.* (2) If the numbers in a sequence s_1, s_2, . . . , do not grow beyond all bounds but if there exists a fixed number M such that $s_n \leq M$, then the sequence is called *bounded above.*

THEOREM II. *A sequence which is both monotonic and bounded is convergent, and*

$$s_n \leq \lim_{n\to\infty} s_n = s .$$

Theorem I is only the general formulation of a situation which we discussed fully above. The reasoning used in establishing the "concrete result," that is, the convergence of

$$s_n = \left(1 + \frac{1}{n}\right)^n \quad \text{and} \quad t_n = \left(1 - \frac{1}{n}\right)^{-n}$$

(or of s_n and t_n where these were the sides* of the *n*-gons inscribed in and circumscribed about a circle), had nothing to do with the particular situations; hence it is not necessary to repeat it here to prove the general case.

* [The German original has "area," which is a slip.—EDITOR.]

On the other hand, Theorem II requires a proof. This will be our first general proof to be given without preparation on particular examples. But it is time that we try our hand at it. First some preliminary remarks: The sequence $\sqrt{1}, \sqrt{2}, \sqrt{3}, \sqrt{4}, \ldots$, is monotonic but is not bounded; the sequence $2\frac{1}{2}, \frac{1}{2}$, $2\frac{1}{3}, \frac{2}{3}, 2\frac{1}{4}, \frac{3}{4}, \ldots$, is bounded above but is not monotonic. This is merely to show what we mean by requiring the sequence to be both monotonic and bounded.

Since the sequence is to be bounded, there is an M such that all $s_n \leq M$. This M need not be an integer. If it is not an integer, we take the nearest integer N to the right of M. Then $s_n \leq N$. On the other hand, let m be the nearest integer to the left of s_1. In case s_1 should by chance be itself an integer, then $m = s_1$. Now the whole sequence lies between m and N (Fig. 16):

$$m \leq s_1 \leq s_2 \leq \ldots \leq M \leq N .$$

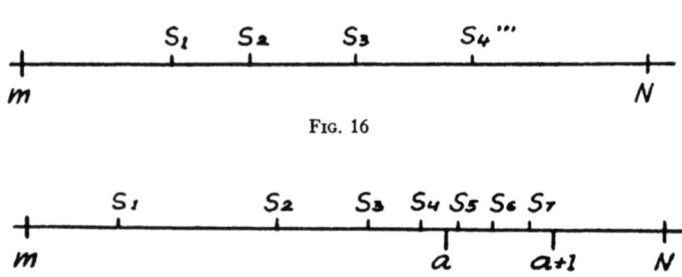

FIG. 16

FIG. 17

The interval between the two integers m and N then consists of $N - m$ intervals of unit length, from m to $m + 1$, from $m + 1$ to $m + 2$, $\ldots$, from $N - 1$ to N. We include the left end point in the interval but not the right one and, hence, write $[m, m + 1)$. The s_n, of course, need not reach into $[N - 1, N)$; all we know is that they are all to the left of M but not how far to the left. But, if they fail to reach into the interval $[N - 1, N)$, there must be a right-most among the $N - m$ intervals into which they still reach, say, the interval $[a, a + 1)$ (Fig. 17). That means that all $s_n < a + 1$, but not all $s_n < a$, since otherwise $[a, a + 1)$ would be free of numbers s_n. On the other hand, so all s_n have to lie in $[a, a + 1)$; some of the first may lie to the left of a. But some one s_n must lie in $[a, a + 1)$; and hence all later ones must likewise be in $[a, a + 1)$ because, on the one hand, the sequence is monotonic, and at the same time all s_n remain below $a + 1$. If we present the s_n as decimal fractions, all except a certain number of early ones begin with a.

Next we divide $[a, a + 1)$ into ten equal parts. Then the first subinterval reaches from the integer a to $a + \frac{1}{10}$ or, decimally written, from a to $a.1$; the second, from $a.1$ to $a.2$, etc. Again it is not necessary for any s_n to fall in the tenth subinterval $[a.9, a + 1)$. But among the ten subintervals there is bound to be one farthest to the right which still contains numbers s_n though perhaps not all the s_n—surely not those that were to the left of a, and even perhaps some others too. But beginning with some s_n all the subsequent ones will then fall in this rightmost interval

$$\left[a + \frac{b}{10}, \quad a + \frac{b+1}{10} \right).$$

Written as decimals, they will therefore all begin with $a.b$, that is, all be $a.b \ldots$.

This procedure we continue. It shows that, from some possibly later one onward, all s_n will begin with $a.bc \ldots$, from a still later one with $a.bcd \ldots$, etc. In this manner we obtain a definite infinite decimal fraction $\rho = a.bcd \ldots$. For all s_n we then have $s_n < a + 1$; but also $s_n < a.(b+1) \leq \rho + \frac{1}{10}$; also $s_n < a.b(c+1) \leq \rho + 1/10^2$; etc. All s_n are thus less than or equal to ρ. On the other hand, all s_n from a certain n onward begin with $a.b$; farther on, with $a.bc$; etc. That is, they come indefinitely closer to ρ. They "consolidate themselves toward ρ," as we used to say; they "converge to ρ," as we say now.

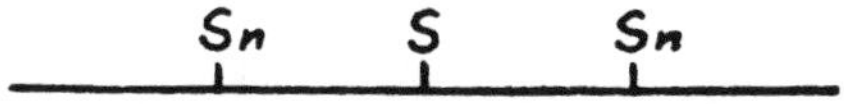

FIG. 18

The proofs of the following theorems will be much easier than this one. For the sake of convenience we shall, however, introduce a new concept which will often be handy—the concept of *absolute value*. We have repeatedly had to say "s_n differs from s by less than ϵ." By this cautious expression we avoided committing ouselves as to whether s_n lies to the left or to the right of s, that is, whether $s - s_n \leq \epsilon$ or $s_n - s \leq \epsilon$ (Fig. 18). In other words, we abstained from using a formula so as to evade this dual possibility which would require our writing two formulas to choose from as the case might require. To do without a formula, however, and thus to lose the advantage of routine manipulation, would eventually become very burdensome. Hence we shall introduce a new symbol to help us out of this difficulty.

Let $|a|$, the *absolute value* of a, mean the value of a without regard for the sign.*

$$|-7| = 7, \quad |7| = 7, \quad |-\pi| = \pi, \quad |-\tfrac{1}{2}| = \tfrac{1}{2}, \quad |0| = 0.$$

For this symbol we readily obtain the following theorems:

1. $|a + b| \leq |a| + |b|$;

2. $|ab| = |a| \, |b|$;

3. $|a - b| \geq |a| - |b|$;

* [The symbol $|a|$ is that one of the numbers a and $-a$ which is not negative.—EDITOR.]

4. $|a_1 + \ldots + a_n| \leq |a_1| + \ldots + |a_n|$;

5. $|a_1 a_2 \ldots a_n| = |a_1| \, |a_2| \, \ldots \, |a_n|$;

6. $|a^n| = |a|^n$;

7. $|a - b| = |b - a|$.

After these preparations we shall next prove some simple theorems concerning convergence.

THEOREM 1. *If two sequences* $s_1, s_2, \ldots,$ *and* $t_1, t_2, \ldots,$ *are both convergent, then the sequence* $s_1 + t_1, s_2 + t_2, \ldots,$ *is also convergent, and*

$$\lim_{n \to \infty} (s_n + t_n) = \lim_{n \to \infty} s_n + \lim_{n \to \infty} t_n .$$

Proof. The assumption that s_n converges to s means that, for any $\epsilon > 0$ we can find a number p such that $|s_n - s| \leq \epsilon$ for $n \geq p$. Similarly, for t_n to converge to t means that a number q can be found such that $|t_n - t| \leq \epsilon$ for $n \geq q$. The assertion that $s_n + t_n$ converges to $s + t$ means that, for any $\epsilon > 0$, a number r can be found such that $|(s_n + t_n) - (s + t)| \leq \epsilon$ for $n \geq r$. It is the existence of such an r which must be proved. For this purpose we use instead of the ϵ of the assumption the value $\epsilon/2$; that is, we determine the p so that $|s_n - s| \leq (\epsilon/2)$ for $n \geq p$ and the q so that $|t_n - t| \leq (\epsilon/2)$ for $n \geq q$. Then we have

$$|(s_n + t_n) - (s + t)| = |(s_n - s) + (t_n - t)| \leq |s_n - s|$$
$$+ |t_n - t| \leq \frac{\epsilon}{2} + \frac{\epsilon}{2} = \epsilon.$$

where n must satisfy both conditions $n \geq p$ and $n \geq q$. Hence we obtain the r in question by making it equal to the greater of the two numbers p and q. This proves our theorem.

THEOREM 2. *Every convergent sequence is bounded.*

Proof. Let s be the limit of the sequence s_n. We assign a value to ϵ, say, 1. Then there is a p such that $|s_n - s| \leq \epsilon$ for all $n \geq p$. All s_n for $n \geq p$ thus lie between $s - 1$ and $s + 1$, and at most $p - 1$ terms of the sequence lie outside the interval $[s - 1, s + 1]$. Hence it is possible to choose $M > 0$ such that both the interval $[s - 1, s + 1]$ and the finite number of terms of the sequence outside the interval will all lie inside the interval $[-M, +M]$. This, however, means $|s_n| \leq M$ for all n; that is, the sequence is bounded.

THEOREM 3. *If two sequences* $s_1, s_2, \ldots,$ *and* $t_1, t_2, \ldots,$ *are both convergent, then the sequence* $s_1 t_1, s_2 t_2, \ldots,$ *is also convergent, and*

$$\lim_{n \to \infty} s_n t_n = \lim_{n \to \infty} s_n \cdot \lim_{n \to \infty} t_n .$$

Proof. We begin by deriving an inequality for the expression $|st - s_n t_n|$ which will be helpful in the proof. For all n we have

$$
\begin{aligned}
|st - s_n t_n| &= |(st - s_n t) + (s_n t - s_n t_n)| \\
&= |t(s - s_n) + s_n(t - t_n)| \\
&\leq |t(s - s_n)| + |s_n(t - t_n)| \\
&\leq |t|\,|s - s_n| + |s_n|\,|t - t_n|.
\end{aligned}
$$

What we have to prove now is that for any preassigned $\epsilon > 0$ an r can be found such that $|st - s_n t_n| \leq \epsilon$ for any $n \geq r$. Now let ϵ be given. From Theorem 2 we know that the convergent sequence s_n is bounded—that there exists a bound M such that, for all n, $|s_n| \leq M$. The same is true for the sequence t_n. We choose M positive and large enough to be a bound for both sequences. Again, because of the convergence of the two sequences, a number p can be determined such that $|s - s_n| \leq (\epsilon/2M)$ for all $n \geq p$ and a number q such that $|t - t_n| \leq (\epsilon/2M)$ for all $n \geq q$. Let r be the greater of these two numbers. Then we have, for $n \geq r$,

$$
|st - s_n t_n| \leq M\,\frac{\epsilon}{2M} + M\,\frac{\epsilon}{2M} = \frac{\epsilon}{2} + \frac{\epsilon}{2} = \epsilon.
$$

This is what was to be proved.

Though this proof may look ingenious, it really is not. It contains no clever ideas; everything is mere routine. The extras that appeared on the stage somewhat unexpectedly were in fact all manipulated from behind the scene, so that in the end everything should come out neat and smooth.

THEOREM 4. *If* $s_1 \neq 0$, $s_2 \neq 0$, . . . , *and if* $\lim_{n \to \infty} s_n = s$ *exists and* $s \neq 0$, *then* $\lim_{n \to \infty} 1/s_n$ *also exists and is equal to* $1/s$.

Proof. First, the sequence $1/s_1$, $1/s_2$, . . . , is bounded. For if, for each M there were some $|1/s_n| > M$, then $|s_n| < (1/M)$; that is, for every $\epsilon = 1/M$ there would be some s_n such that $|s_n| < \epsilon$. That, however, would contradict the hypothesis

$$
\lim_{n \to \infty} s_n = s \neq 0.
$$

Hence there is some M such that, for all n, $|1/s_n| \leq M$. Next,

$$
\left| \frac{1}{s} - \frac{1}{s_n} \right| = \left| \frac{s_n - s}{s\,s_n} \right| = \frac{1}{|s|}\,|s_n - s|\,\frac{1}{|s_n|} \leq \frac{1}{|s|}\,|s_n - s|\,M.
$$

Because of the hypothesis $\lim_{n \to \infty} s_n = s$, for increasing n, $|s_n - s|$ will approach zero. Hence we can choose a number p such that, for $n \geq p$, we have

$$
|s_n - s| \leq \frac{\epsilon\,|s|}{M}.
$$

Then $|(1/s) - (1/s_n)| \leq \epsilon$ for $n \geq p$; that is, $1/s_n$ converges to $1/s$, as was to be proved.

THEOREM 5. *If* $s_n \leq t_n$ *for every* n, *and if both sequences converge, then*

$$s = \lim_{n \to \infty} s_n \leq \lim_{n \to \infty} t_n = t .$$

Proof. Suppose $s > t$. Now the following holds:

$$t - s = (t - t_n) + (t_n - s_n) + (s_n - s) ;$$

therefore,

$$(t - s) - (t_n - s_n) = (t - t_n) + (s_n - s) .$$

Let $\epsilon = |t - s|/2$. According to the hypothesis, we can find two numbers p and q such that for $n \geq p$ we shall have $|s_n - s| \leq (\epsilon/2)$, and for $n \geq q$ we shall have $|t - t_n| \leq (\epsilon/2)$. Hence, if r is taken to be the greater of the two numbers p and q, for $n \geq r$ we shall have

$$|(t - s) - (t_n - s_n)| \leq \epsilon = \frac{|t - s|}{2}. \tag{9.1}$$

If $s > t$, then $t - s$ would be negative. Since, according to hypothesis, $-(t_n - s_n)$ is negative or zero, we would then have

$$|(t - s) - (t_n - s_n)| \geq |t - s| . \tag{9.2}$$

But (9.1) and (9.2) together show that $|t - s| \leq 0$, from which it follows that $t - s = 0$. The assumption $s > t$ is therefore untenable; hence $s \leq t$, as was to be proved.*

Now for an application. What is $\lim_{n \to \infty} \sqrt[n]{n}$? In the beginning the sequence acts rather strangely:

$$\sqrt[1]{1} = 1 ; \qquad \sqrt[2]{2} = 1.4142 \ldots ; \qquad \sqrt[3]{3} = 1.4422 \ldots ;$$

$$\sqrt[4]{4} = \sqrt[2]{2} = 1.4142 \ldots ;$$

that is, there is an increase up to $\sqrt[3]{3}$, after which we seem to be going downhill.

We are indeed going downhill from there on. For if we ask for what values of n one has $\sqrt[n]{n} > \sqrt[n+1]{n+1}$, we see that for such n the following inequality must hold:

$$n^{n+1} > (n+1)^n, \quad \text{or} \quad n > \frac{(n+1)^n}{n^n} = \left(\frac{n+1}{n}\right)^n .$$

But $\lim_{n \to \infty} [1 + (1/n)]^n = e = 2.718 \ldots$, and $[n + (1/n)]^n$ approaches e from below. Hence we can conclude that, if $n > e$, then

$$\left(1 + \frac{1}{n}\right)^n < e < n, \qquad \left(\frac{n+1}{n}\right)^n < n, \qquad (n+1)^n < n^{n+1},$$

* [The proof as a whole has been somewhat rearranged from the German original.—EDITOR.]

and, finally,

$$\sqrt[n+1]{n+1} < \sqrt[n]{n} .$$

Hence, beginning with $n = 3$, the value of $\sqrt[n]{n}$ does indeed decline. That, of course, does not mean that it approaches zero as a limit. In fact, $\sqrt[n]{n} \geq 1$ always. We can therefore apply Theorem II* to conclude that $\sqrt[n]{n}$ approaches a limit. Let us call this limit X; we already know that $X \geq 1$. Now we consider a partial sequence of $s_n = \sqrt[n]{n}$, namely, $t_n = \sqrt[2n]{2n}$, which means $t_1 = s_2, t_2 = s_4, t_3 = s_6 \dots$. This sequence therefore converges also to X. But $\sqrt[2n]{2n} = \sqrt[2n]{2}\sqrt[2n]{n} = \sqrt[n]{\sqrt{2}} \cdot \sqrt[n]{\sqrt[n]{n}}$. With increasing n, $\sqrt[n]{\sqrt{2}}$ approaches 1 (see Exercise No. 3). Hence the right side goes to $1\sqrt{X} = \sqrt{X}$,† while the left side approaches X. We thus obtain for the unknown limit X the equation $X = \sqrt{X}$; hence, $X^2 = X$, $X(X - 1) = 0$, and, since $X \geq 1$, $X = 1$. We have thus proved that

$$\lim_{n \to \infty} \sqrt[n]{n} = 1 .$$

10. INFINITE SERIES

Beginners find it sometimes difficult to distinguish between "sequences" and "series." To make it easier for them, we insert this section on series right after the one on sequences, even though their full treatment will be given only in chapter iv.‡

We have already encountered a convergent series. We said:

$$\tfrac{1}{2} + \tfrac{1}{4} + \tfrac{1}{8} + \dots = 1 .$$

---

* [Theorem II (p. 33) as it stands is not applicable, but the needed corollary for a sequence which is decreasing and bounded below can be readily derived.—EDITOR.]

† [It is assumed here that if s_n is a sequence with $\lim_{n \to \infty} s_n = X$, then $\sqrt{s_n}$ is a sequence with $\lim_{n \to \infty} \sqrt{s_n} = \sqrt{X}$. To prove this, note that

$$| (\sqrt{s_n} - \sqrt{X})(\sqrt{s_n} + \sqrt{X}) | = | s_n - X | .$$

Now $\sqrt{s_n}$ is bounded by a positive number M, since s_n is bounded; hence

$$| \sqrt{s_n} - \sqrt{X} | < | s_n - X | / (\sqrt{X} + M) .$$

By definition of the limit for s_n, there exists an integer p such that

$$| s_n - X | < (\sqrt{X} + M)\epsilon$$

for all $n \geq p$ and some preassigned $\epsilon > 0$. Hence

$$| \sqrt{s_n} - \sqrt{X} | < \epsilon$$

for all $n \geq p$.—E. T.]

‡ [Not included in this volume.—EDITOR.]

Why? What was it that converged? When we said, "This series converges," we meant in fact this:

$$\tfrac{1}{2} \qquad\quad = \tfrac{1}{2}$$

$$\tfrac{1}{2} + \tfrac{1}{4} \qquad = \tfrac{3}{4}$$

$$\tfrac{1}{2} + \tfrac{1}{4} + \tfrac{1}{8} = \tfrac{7}{8}$$

. . .

and the values $\tfrac{1}{2}, \tfrac{3}{4}, \tfrac{7}{8} \ldots$, approach ever closer to 1. We did not add up all terms of the series; we added the first n of them and then let n increase. On this process we shall base our general definition:

DEFINITION. *The infinite series* $a_1 + a_2 + \ldots$ *is called convergent if the sequence of its "partial sums"* $s_1 = a_1$, $s_2 = a_1 + a_2$, $\ldots$, $s_n = a_1 + a_2 + \ldots + a_n$ *is convergent; and the limit* $s = \lim\limits_{n \to \infty} s_n = \lim\limits_{n \to \infty} (a_1 + \ldots + a_n)$ *is called "the sum" of the series:* $a_1 + a_2 + \ldots = s$.

The *series* $1 + \tfrac{1}{2} + \tfrac{1}{4} + \ldots$ is thus something quite different from the *sequence* $1, \tfrac{1}{2}, \tfrac{1}{4}, \ldots$. Both are convergent. But the series has the sum 2; the sequence has the limit 0. It is different with the series and sequence

$$1 + \frac{1}{\sqrt{2}} + \frac{1}{\sqrt{3}} + \cdots, \qquad \text{and} \qquad 1, \; \frac{1}{\sqrt{2}}, \; \frac{1}{\sqrt{3}}, \; \cdots .$$

The sequence has again the limit zero. The series, however, does not approach a limit at all. This can at once be seen from the inequality

$$1 + \frac{1}{\sqrt{2}} + \cdots + \frac{1}{\sqrt{n}} \geq \frac{1}{\sqrt{n}} + \frac{1}{\sqrt{n}} + \cdots + \frac{1}{\sqrt{n}} = \frac{n}{\sqrt{n}} = \sqrt{n}.$$

Since $\sqrt{n}$ increases without limit with n, each partial sum s_n, which is always greater than $\sqrt{n}$, does likewise. Hence here we have a case where the series is "divergent," that is, not convergent, even though the sequence of its terms approaches zero. For the terms of a series to approach zero is thus not enough to assure the convergence of the series; it is not a "sufficient condition." But it is a "necessary condition," as will be shown next.

THEOREM 1. *If* $a_1 + a_2 + \ldots$ *is convergent, then* $\lim\limits_{n \to \infty} a_n = 0$.

Proof. Let p be a number such that for $n \geq p$ the difference $|s_n - s| \leq (\epsilon/2)$, where ϵ is some preassigned positive number. Then, for $n \geq p + 1$ we have $|a_n| = |s_n - s_{n-1}| = |(s_n - s) - (s_{n-1} - s)| \leq |s - s_n| + |s - s_{n-1}| \leq (\epsilon/2) + (\epsilon/2) = \epsilon$. But this means that $\lim\limits_{n \to \infty} a_n = 0$.

Let us point out once more that we have here a theorem whose "converse" is not true. If $a_1 + a_2 + \ldots$ converges, then $\lim\limits_{n \to \infty} a_n = 0$; but, if $\lim\limits_{n \to \infty} a_n = 0$, $a_1 + a_2 + \ldots$ need not be convergent. So $\lim\limits_{n \to \infty} a_n = 0$ is a *necessary condition* for the convergence of the series but not a *sufficient condition*. If it rains, then the street is wet; but, if the street is wet, it need not have rained. That seems obvious to anybody. It is something else for everybody to avoid improper conversion of statements. Even sharp-minded legal experts have been known to confuse a statement and its converse. The beginner in mathematics needs some training not to slip up on this point.

The convergence of series was not a new definition but was based on the convergence of sequences. Hence all theorems concerning convergence of sequences can be applied to series.

THEOREM I. *If the series* $b_1 - b_2 + b_3 - b_4 \pm \ldots$ *satisfies the following three conditions*

1. $b_n \geq 0$,
2. $b_1 \geq b_2 \geq b_3 \geq \ldots$,
3. $\lim\limits_{n \to \infty} b_n = 0$,

then the series is convergent.

Proof. Consider the two sequences:

$$s_1 = b_1 \qquad\qquad s_2 = b_1 - b_2$$

$$s_3 = b_1 - b_2 + b_3 \qquad\qquad s_4 = b_1 - b_2 + b_3 - b_4$$

$$s_5 = b_1 - b_2 + b_3 - b_4 + b_5 \qquad s_6 = b_1 - b_2 + b_3 - b_4 + b_5 - b_6$$

. . .

For these the following four statements can be made:

1. $s_2 \leq s_4 \leq s_6 \leq \ldots$, since $s_4 = s_2 + (b_3 - b_4)$, and $b_3 - b_4 \geq 0$, etc.
2. $s_1 \geq s_3 \geq s_5 \geq \ldots$, since $s_3 = s_1 - (b_2 - b_3)$, and $b_2 - b_3 \geq 0$, etc.
3. $s_1 \geq s_2, s_3 \geq s_4, s_5 \geq s_6, \ldots$.
4. The sequence $s_1 - s_2 = b_2, s_3 - s_4 = b_4, s_5 - s_6 = b_6, \ldots$, converges to zero, according to Condition 3 stated in the theorem.

Thus all assumptions of Theorem I on sequences are satisfied, and, therefore, the series is proved convergent.

THEOREM II. *If all* $b_n \geq 0$, *and if* $b_1 + \ldots + b_n$ *remains less than a bound* M, *which is independent of* n, *then the series* $b_1 + b_2 + \ldots$ *converges, and its sum* b *is less than or equal to* M.

Proof. This follows at once from the proof of Theorem II [p. 33] in the previous section.

Immediate consequences are the following:

THEOREM 2. *If* $a_1 + a_2 + \ldots = s$, *and* $b_1 + b_2 + \ldots = t$ *are convergent, then the series* $(a_1 + b_1) + (a_2 + b_2) + \ldots$ *is convergent, and its sum is* $s + t$.

THEOREM 3. *If* $a_1 + a_2 + \ldots = s$ *is convergent, then the series* $\rho a_1 + \rho a_2 +$ $\ldots$ *is also convergent, and its sum is* ρs.

THEOREM 4. *If* $0 \le a_n \le b_n$, *and if* $b_1 + b_2 + \ldots = t$ *is convergent, then* $a_1 + a_2 + \ldots$ *is convergent with sum* s, *and* $s \le t$.

Now let us give an illustration.

$$\frac{1}{1 \cdot 2} + \frac{1}{2 \cdot 3} + \ldots + \frac{1}{(n-1)n} = (1 - \tfrac{1}{2}) + (\tfrac{1}{2} - \tfrac{1}{3})$$
$$+ \ldots + \left(\frac{1}{n-1} - \frac{1}{n}\right) = 1 - \frac{1}{n}.$$

The infinite series $1/1 \cdot 2 + 1/2 \cdot 3 + \ldots$ is thus seen to converge to the limit 1. Next, the terms of the series $1/2 \cdot 2 + 1/3 \cdot 3 + 1/4 \cdot 4$ are, term for term, less than those of the first series; hence, the second series converges to a limit s less than or equal to 1. From this we can say about the series $1 + \tfrac{1}{4} + \tfrac{1}{9} + \ldots$ that it converges to a limit less than or equal to $1 + 1 = 2$.

2

THE DEFINITE INTEGRAL

11. THE QUADRATURE OF THE PARABOLA BY ARCHIMEDES

In the Preface to his treatise on the quadrature of the parabola[20] Archimedes writes: "Many mathematicians have endeavored to square the circle, the ellipse, or the segment of a circle of an ellipse or of a hyperbola; and the edifice of lemmas which they erected for this purpose has generally been found open to objection. No one, however, seems to have thought of attempting the quadrature of the segment of a parabola, which is precisely the one which can be carried out."

For the Greeks ellipse, hyperbola, and parabola were originally defined as "conic sections," that is, intersections of a "right circular cone" with a plane (Fig. 19). A "right circular cone" is generated in the following way: At the center of a circle M we erect the perpendicular to the plane of the circle. From a point O on this perpendicular the pencil of rays to the points P_1, P_2, ..., of the circle forms a surface, the so-called right circular cone (Fig. 20). We can extend the rays OP_1, OP_2, ... Then the circular cone extends indefinitely downward. But we can also extend the rays upward beyond O, so that they are full straight lines. All such extensions are regarded as belonging to the "right circular cone," which thus consists, so to speak, of two infinitely extended sugar loaves* fastened together at their peaks. Now the question is to cut this object with a plane.

If we place the plane parallel to the original circle, the intersection is readily seen to be always a circle. If we incline the plane somewhat, the intersection becomes narrowed and longer. This curve is an ellipse. If we incline the plane a great deal—for example, perpendicular to the plane of the circle—the intersection will consist of two parts, one on each of the two sugar loaves. This curve is a hyperbola; it looks like the one sketched in Figure 21.

Between these two situations—the small and the almost perpendicular inclination—lies a transitional one, as indicated in Figure 22. If, namely, the plane lies parallel to a generator of the cone, we obtain in the lower section a curve no longer closed but continuing indefinitely downward, while the upper sugar loaf is not yet touched. This curve of intersection is the parabola.

Of the theory of the conic sections much was known at the time of Archi-

43

*A sugar loaf is a block of sugar molded into a cone and rounded at the top. In the 1930s, this was an example from common experience of a shape similar to a paraboloid or one sheet of a hyperboloid.

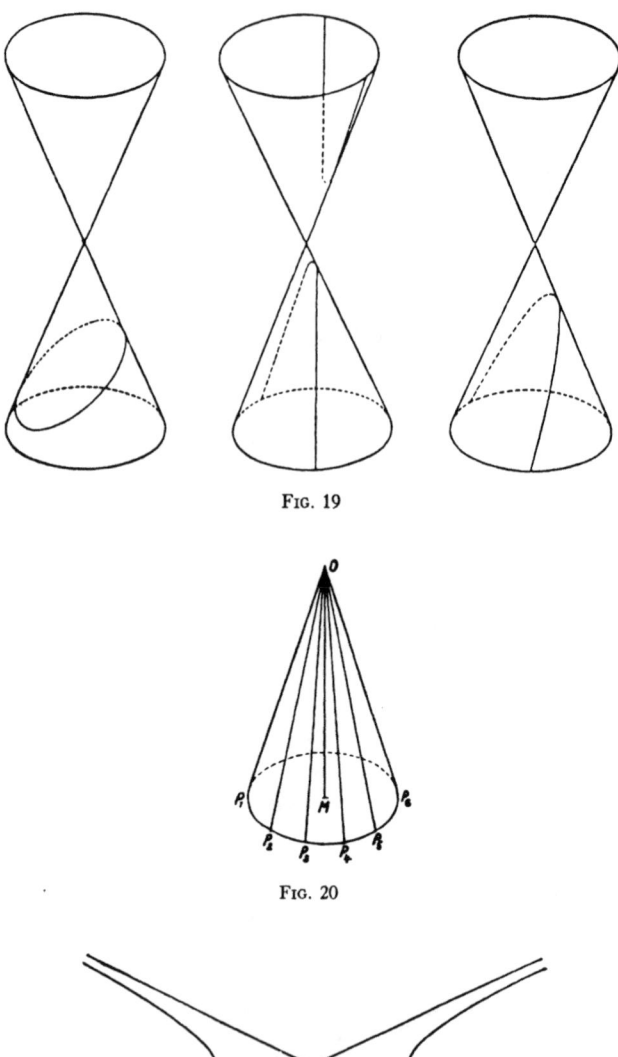

FIG. 19

FIG. 20

FIG. 21

medes (250 B.C.). This theory is often credited to Euclid (300 B.C.), whose famous textbook, however, presented mostly other people's discoveries. His book on the conic sections is lost. It was superseded by the work of Apollonius (200 B.C.).[21] The theory of the conic sections began by replacing the readily visualized spatial definition given above by one using relations in a plane only.

Our interest here is confined to the parabola. Its definition can be given in the following form, which is the one used by Archimedes in his treatise: Take a parallelogram $EE'F'F$ (Fig. 23); let O be the midpoint of EE', and A that

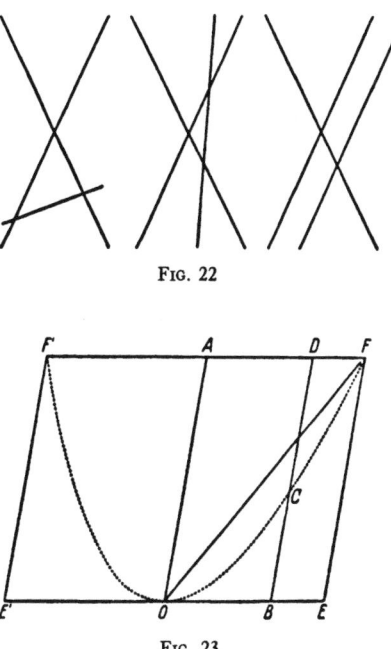

FIG. 22

FIG. 23

of FF'. Let B be an arbitrary point on EE', and BD the parallel to EF through B. Now we define on BD a point C such that

$$BC:EF = BO^2:EO^2 . \qquad (11.1)$$

Let us see for a moment what this means. If we had required the point C to satisfy the proportion

$$BC:EF = BO:EO ,$$

the answer would be very simple: In accordance with the theory of similar triangles (Fig. 24), C would be the intersection of BD with OF. However, $(BO/EO)^2$ is different from BO/EO. Hence C is not the point of intersection with OF but some other point of BD.

If point B now travels from E to E', C will likewise travel. The locus of all these points C is the parabola (dotted curve in Fig. 25). The proof that this locus is also the section of a right circular cone does not concern us here. For us the parabola is now this locus described above; that is, it is defined by the proportion (11.1).

Archimedes gave two proofs for his quadrature of the segment of the parabola. The second proof uses nothing but the definition just stated. The first

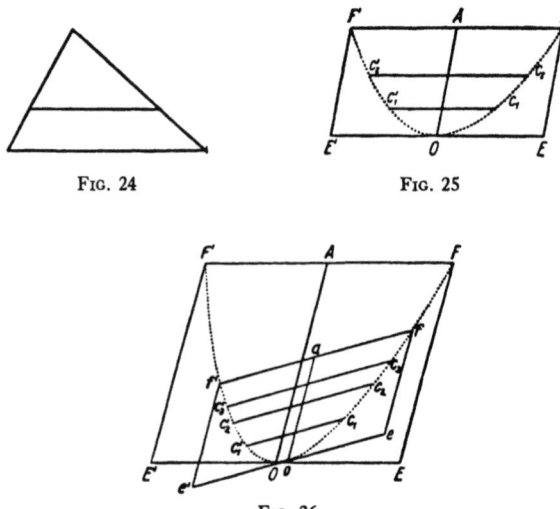

FIG. 24   FIG. 25

FIG. 26

proof, however, makes use also of three theorems concerning parabolas. I state them here without proof (as did also Archimedes):

1. If the chords C_1C_1', C_2C_2', ..., are parallel to EE', their midpoints lie on OA (Fig. 25).

2. If ff' is any chord of the parabola, then the midpoints of all chords parallel to ff' lie on a line oa parallel to OA, where o is the point of tangency of the tangent parallel to ff' (Fig. 26).*

3. In this case a proportion corresponding to (11.1) holds:

$$bc:ef = bo^2:eo^2 .$$

First proof. As a starting point we take the definition of the parabola. Choosing B (Fig. 27) to be the midpoint of the segment OE, and B' to be the midpoint of OE', we have $BO:EO = 1:2$, and hence $BO^2:EO^2 = 1:4$. By the definition (11.1) of the parabola, $BC:EF = 1:4$. But BD being parallel to EF, we

---

* [The relationship of b to the points c, e, f, and o in Fig. 26 corresponds to that of B to the points C, E, F, and O in Fig. 23.—EDITOR.]

have $BD = EF$. Now G is immediately seen to be the midpoint of BD. But $BC = \frac{1}{4}BD$, and hence C is the midpoint of BG; therefore,

$$DG = 2GC. \tag{11.2}$$

Now we repeatedly use the theorem that two triangles (ABO and BCO in Fig. 28) with equal base and equal height have equal areas. Observe that (Fig. 29) the triangle ABO has height equal to and base equal to twice that of BCO;

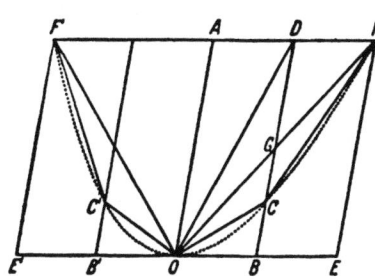

Fig. 27

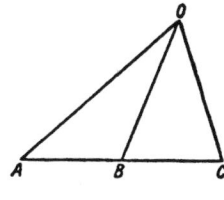

Fig. 28

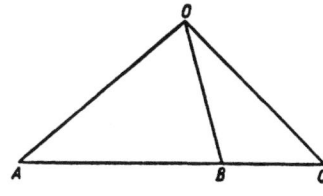

Fig. 29

so its area is twice that of BCO. Applying these relations to the right half of our parabola (Fig. 27), we have

$$DGF = 2GCF$$

and

$$DGO = 2GCO.$$

Hence, by adding these areas,

$$FOD = 2FOC.$$

Further

$$FOA = 2FOD,$$

and hence, because of the preceding statement,

$$FOA = 4FOC;$$

therefore,

$$FOF' = 2FOA = 8FOC. \tag{11.3}$$

For exactly the same reasons, applied to the left half of our parabola (Fig. 27), we have

$$FOF' = 8F'OC'. \qquad (11.4)$$

From (11.3) and (11.4)

$$FOC = F'OC',$$

so that

$$FOF' = 4(FOC + F'OC'),$$

or

$$FOC + F'OC' = \tfrac{1}{4}FOF'. \qquad (11.5)$$

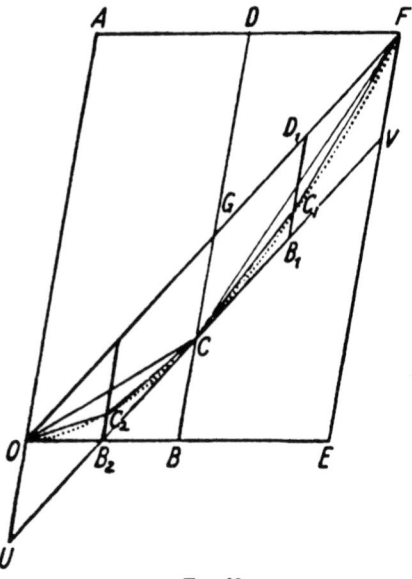

Fig. 30

In Figure 30 we show only the right half of Figure 27; we apply to it Theorem 3, the last of the theorems with which we started. Take OF to be the chord $f'f$ (notations as in Fig. 26): G has the role of a, C the role of o, and UV the role of $e'e$. Now we bisect $CV = oe$ in B_1, $CU = oe'$ in B_2. Then—applying the theorem—we obtain $B_1C_1 : VF = B_1C^2 : VC^2$. Now, since $B_1C = \tfrac{1}{2}VC$ and $B_1D_1 = VF$, the parabola divides B_1D_1 in C_1 in the ratio 1:4.

We can show, arguing as before, that

$$fof' = 4(foc + f'oc');$$

hence, since $fof' = FOC$, $foc = FCC_1$ and $f'oc' = OCC_2$,

$$FOC = 4(FCC_1 + OCC_2),$$

or

$$FCC_1 + OCC_2 = \tfrac{1}{4}FOC.$$

Now comes the point! Our goal is to get an expression for the area of the segment of the parabola cut off by the chord FF' (Fig. 27). Triangle FOF' is part of this segment—even a considerable part. If to FOF' we add the two triangles FOC and $F'OC'$, we have the figure $FCOC'F'$, bounded completely by straight-line segments. It, too, is a part of the parabolic segment under consideration, a considerably larger part than the triangle FOF'. On the other hand, from (11.5) we obtain

$$FCOC'F' = FOF' + (FOC + F'OC') = FOF' + \tfrac{1}{4}FOF' .$$

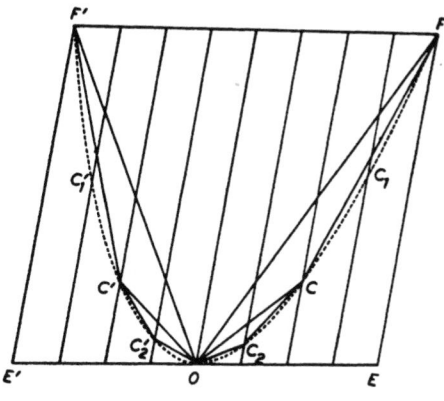

FIG. 30A

From here it follows further that, for the polygon $FC_1CC_2OC_2'C'C_1'F'$ (Fig. 30A),

$$FC_1CC_2OC_2'C'C_1'F' = FOF' + (FOC + F'OC') + [(FC_1C + CC_2O)$$

$$+ (OC_2'C' + C'C_1'F')]$$

$$= FOF' + \tfrac{1}{4}FOF' + (\tfrac{1}{4}FOC + \tfrac{1}{4}F'OC')$$

$$= FOF'(1 + \tfrac{1}{4} + \tfrac{1}{16}) .$$

If we keep repeating this procedure, the area of the polygon which always is contained in that segment of the parabola approaches the latter more and more. On the right-hand side we then have

$$FOF' \left(1 + \frac{1}{4} + \frac{1}{4^2} + \frac{1}{4^3} + \dots \right).$$

Since, for $x < 1$, the value of a geometric series is

$$1 + x + x^2 + \dots = \frac{1}{1-x},$$

we obtain, for $x = \frac{1}{4}$,

$$\frac{1}{1 - \frac{1}{4}} = \frac{1}{\frac{3}{4}} = \frac{4}{3}.$$

That is,

$$\text{Segment of parabola} = \tfrac{4}{3}FOF' \ !$$

The areas of the segment of the parabola and of the triangle FOF' are thus commensurable to each other, and it is an easy matter of elementary geometry to construct a square equal to $\tfrac{4}{3}FOF'$ with compass and ruler. Thus the segment of the parabola has been "squared."

Let us point out the importance which the result of Archimedes had for his own time, even though its enormous significance for the subsequent development will appear only further in the course of this chapter. It constitutes a tremendous discovery! Let us remember that Hippocrates had squared a certain

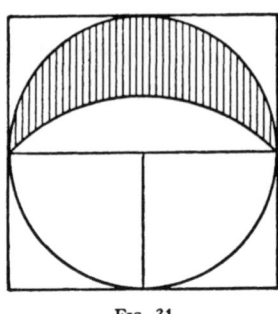

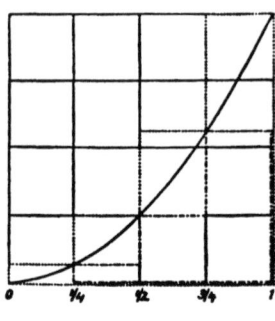

FIG. 31   FIG. 32

crescent (Fig. 31) and a few similar ones and thus demonstrated that there exist curvilinear figures which can be "squared."

After that, from 450 to 260 B.C., some of the greatest minds, and many lesser ones as well, had tried to square other curvilinear figures, most of all the full circle, or the ellipse, or segments of the hyperbola—always in vain! And here comes young Archimedes—for it seems that this was his first discovery—and shows that any segment of a parabola is equal to $\tfrac{4}{3}$ of a certain inscribed triangle. This, in itself, is already an epoch-making event. Its full significance we shall come to appreciate only gradually.

Second proof. This time we begin with a parabolic segment contained, not in a parallelogram, but in a rectangle—in fact, in a rectangle whose base equals twice its height.* We show only the right half of the rectangle with the parabolic arc in it (Fig. 32). For simplicity we take the height of the rectangle to be 4.

We now prove that the part R of the square which lies under the parabola is equal to $\tfrac{1}{3}$ of the whole square Q. We divide the side of the square into four equal parts and construct the two sets of rectangles shown in the figure; one set is

---

* [These restrictions are not indicated in the German original.—EDITOR.]

bounded by the dotted lines, the other one by the dash-dotted lines. The way these sets are constructed should be obvious from the figure. The figure bounded by the dotted lines contains the area R, whereas the figure bounded by the dash-dotted lines is contained in R, so that R is less than the former but larger than the latter.

Now it is easy to express the areas of the dotted and dash-dotted figures, for both consist of rectangular strips whose width is throughout $\frac{1}{4}$.

$$T_4 = \text{dotted figure} = \tfrac{1}{4}(\tfrac{1}{4})^2 + \tfrac{1}{4}(\tfrac{2}{4})^2 + \tfrac{1}{4}(\tfrac{3}{4})^2 + \tfrac{1}{4}(\tfrac{4}{4})^2\,,$$

$$S_4 = \text{dash-dotted figure} = \tfrac{1}{4}\cdot 0^2 + \tfrac{1}{4}(\tfrac{1}{4})^2 + \tfrac{1}{4}(\tfrac{2}{4})^2 + \tfrac{1}{4}(\tfrac{3}{4})^2\,.$$

Obviously, if we divide the side of the square into n rather than into four parts, we obtain, similarly,

$$T_n = \frac{1}{n}\left(\frac{1}{n}\right)^2 + \frac{1}{n}\left(\frac{2}{n}\right)^2 + \ldots + \frac{1}{n}\left(\frac{n}{n}\right)^2\,,$$

$$S_n = \frac{1}{n}\cdot 0^2 + \frac{1}{n}\left(\frac{1}{n}\right)^2 + \ldots + \frac{1}{n}\left(\frac{n-1}{n}\right)^2\,,$$

and for every n the inequality $S_n < R < T_n$ holds. Moreover, $T_n - S_n$, with all but one term canceling out, is at once seen to be

$$T_n - S_n = \frac{1}{n}\left(\frac{n}{n}\right)^2 = \frac{1}{n}\,.$$

Hence for increasing n the difference between the dotted and the dash-dotted area decreases to zero. This is the computational equivalent of the intuitively seen fact that the dotted figure approaches the area R from above, and the dash-dotted one from below.

But how large is R? Writing T_n and S_n over a common denominator, we have

$$T_n = \frac{1}{n}\left(\frac{1}{n}\right)^2(1^2 + 2^2 + 3^2 + \ldots + n^2) = \frac{1^2 + 2^2 + 3^2 + \ldots + n^2}{n^3}\,,$$

$$S_n = \frac{1}{n}\left(\frac{1}{n}\right)^2[0^2 + 1^2 + 2^2 + \ldots + (n-1)^2] = \frac{0^2 + 1^2 + 2^2 + \ldots (n-1)^2}{n^3}\,.$$

Now we can show that $1^2 + 2^2 + \ldots + n^2 = \tfrac{1}{6}n(n+1)(2n+1)$. (Archimedes proves this in a later treatise on spirals,[22] although the explicit proof is not in the extant papers. Compare also Exercise 2.) Hence

$$T_n = \frac{1}{6}\frac{n(n+1)(2n+1)}{n^3} = \frac{1}{6}\left(1 + \frac{1}{n}\right)\left(2 + \frac{1}{n}\right),$$

and

$$\lim_{n\to\infty} T_n = \tfrac{1}{3}\,;$$

that is,

$$R = \tfrac{1}{3}\,.$$

12. CONTINUATION AFTER 1,880 YEARS

Cavalieri is the first who succeeded, about 1630, in making a further discovery of this kind. If we restate the definition of the parabola given above using present-day notation, we have $y = x^2$. Cavalieri considered the analogous problem for $y = x^3$. This led him to

$$T_n = \frac{1}{n}\left(\frac{1}{n}\right)^3 + \frac{1}{n}\left(\frac{2}{n}\right)^3 + \ldots + \frac{1}{n}\left(\frac{n}{n}\right)^3,$$

$$S_n = \frac{1}{n}\cdot 0^3 + \frac{1}{n}\left(\frac{1}{n}\right)^3 + \ldots + \frac{1}{n}\left(\frac{n-1}{n}\right)^3,$$

and

$$T_n = \frac{1^3 + 2^3 + 3^3 + \ldots + n^3}{n^4}.$$

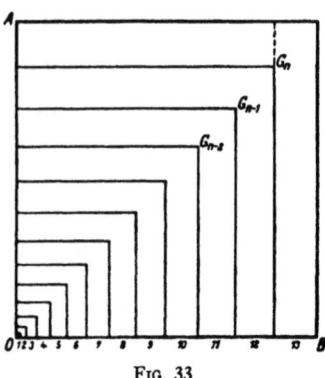

FIG. 33

For the sum of the first n cubes the Arabs[23] knew the formula

$$1^3 + 2^3 + \ldots + n^3 = [\tfrac{1}{2}n(n+1)]^2,$$

for which they gave the following ingenious proof: From the point O (Fig. 33) we lay off successive segments of length 1, 2, 3, etc., and, finally, one of length n, extending up to point B. We next do the same on a line OA perpendicular to OB, so that

$$OA = OB = 1 + 2 + \ldots + n = \tfrac{1}{2}n(n+1).$$

Now we complete the figure to a square, whose area, therefore, is

$$Q = [\tfrac{1}{2}n(n+1)]^2.$$

On the other hand, the large square can be built up of n "angle irons" (called "gnomons" in oriental sacrificial services), so that

$$Q = G_1 + G_2 + \ldots + G_n.$$

But how large is G_n? As shown in the figure, it can be divided by the dotted line into two rectangles, so that

$$G_n = n[\tfrac{1}{2}n(n+1)] + n[\tfrac{1}{2}(n-1)n]$$

$$= \tfrac{1}{2}n^2[(n+1) + (n-1)] = n^3.$$

Thus

$$Q = 1^3 + 2^3 + \ldots + n^3,$$

while above we had

$$Q = [\tfrac{1}{2}n(n+1)]^2.$$

This proves the formula.
For Cavalieri this meant

$$T_n = \frac{[\tfrac{1}{2}n(n+1)]^2}{n^4} = \left[\frac{\tfrac{1}{2}n(n+1)}{n^2}\right]^2 = \left[\frac{1}{2}\left(1+\frac{1}{n}\right)\right]^2,$$

and hence

$$R = \lim_{n\to\infty} T_n = \tfrac{1}{4}.$$

Again a rational number! Again a result like that of Archimedes: Hippocrates, Archimedes, Cavalieri! Cavalieri did not stop here. He studied $y = x^4$ and found $R = \tfrac{1}{5}$; for $y = x^5$ he found $R = \tfrac{1}{6}$; and he continued until $y = x^9$, finding $R = \tfrac{1}{10}$. He would, of course, expect that $y = x^{10}$ would lead to $R = \tfrac{1}{11}$. But for $k = 10$ he found difficulties with the formula

$$1^k + 2^k + \ldots + n^k.$$

This formula meant a new problem for every k, and we must admire Cavalieri for having pushed as far as $k = 9$.[24]

The general result, for arbitrary, k, was obtained by Fermat[25] in one of his last papers around 1650. His method was essentially a generalization of Archimedes' second derivation of the area of a parabolic segment (which is the reason why we gave the two proofs above).

Fermat divided the interval, not into equal, but into unequal intervals, and moreover in such a way that at the very beginning S_n consists of infinitely many parts. We take

$$0 < \rho < 1,$$

where ρ might first be $\tfrac{1}{2}$, then $\tfrac{2}{3}$, then $\tfrac{3}{4}$, in general $\rho = (n-1)/n$, so that, as n increases, ρ approaches 1. For the moment, however, we think of ρ as being fixed, equal to some particular one of these values (Fig. 34). Then we have, for the sum of the rectangles below the curve $y = x^k$,

$$S_\rho = (b - \rho b)(\rho b)^k + (\rho b - \rho^2 b)(\rho^2 b)^k + (\rho^2 b - \rho^3 b)(\rho^3 b)^k + \ldots \infty$$

$$= b^{k+1} \rho^k (1 - \rho)(1 + \rho^{k+1} + \rho^{2k+2} + \rho^{3k+3} + \ldots)$$

$$= b^{k+1} \rho^k (1 - \rho)\frac{1}{1 - \rho^{k+1}} = b^{k+1} \rho^k \frac{1}{(1 - \rho^{k+1})/(1 - \rho)}$$

$$= \frac{b^{k+1} \rho^k}{1 + \rho + \rho^2 + \ldots \rho^k}.$$

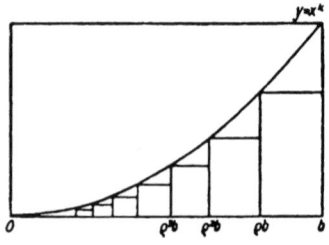

Fig. 34

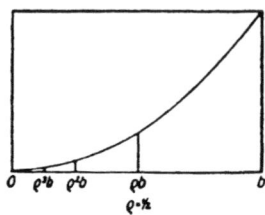

Fig. 35

We used first the theorem on infinite geometric series, then the theorem on finite series, that for the infinite series being, of course, applicable only for $0 < \rho < 1$, which by hypothesis is the cases here.

So far ρ has been fixed. Now we let ρ approach 1. For our subdivisions, which for any ρ have infinitely many partitions, the mesh becomes finer and finer as ρ increases (Fig. 35), and hence S_ρ approaches R. Our formula, however, gives us for any sequence of values of ρ converging to 1, for $\rho \to 1$ as we may briefly say, the limit

$$\lim_{\rho \to 1} S_\rho = \frac{b^{k+1}}{k + 1}$$

which is Cavalieri's rule *for every k.*

For every *k!* Was it necessary in this proof to assume that *k* is an integer?

If we were to choose $k = \frac{3}{2}$ or $k = \frac{1}{2}$, this would not affect our proof until the very last step. That is, we would still have

$$S_\rho = b^{k+1} \rho^k \frac{1}{(1 - \rho^{k+1})/(1 - \rho)}.$$

But, in next applying the theorem on finite geometric series

$$\frac{1 - \rho^{k+1}}{1 - \rho} = 1 + \rho + \rho^2 + \ldots + \rho^k$$

(which for $\rho \to 1$ gave $k + 1$), k was indeed assumed to be an integer. To find

$$\lim_{\rho \to 1} \frac{1 - \rho^{k+1}}{1 - \rho}$$

if k is not an integer—using modern terminology—amounts to the problem of finding the derivative of the function

$$y = x^n \quad \text{at} \quad x = 1, \quad \text{or} \quad \lim_{x \to 1} \frac{1 - x^n}{1 - x} = n;$$

or, by the same token,

$$\lim_{\rho \to 1} \frac{1 - \rho^{k+1}}{1 - \rho} = k + 1.$$

Fermat knew how to handle this problem, for he was at that time in possession of large parts of the differential calculus. Hence he could indeed prove that

$$\lim_{\rho \to 1} S_\rho = \frac{b^{k+1}}{k + 1}$$

for all values of k, except of course for $k + 1 = 0$ or $k = -1$. Strangely, he does not seem to have investigated this case, that is, the area under the curve $y = x^{-1} = 1/x$.

However, at just this time (1647) a most important discovery concerning this case was made by the Jesuit Father Gregorius a Santo Vincentio[26]—a discovery finding its place next to those discoveries of Cavalieri. This discovery is found at the end of his formidably bulky opus on geometry, after hundreds of artful but uninteresting theorems and false statements concerning the quadrature of the circle, which so discredited his opus that those excellent discoveries at the end were almost overlooked.

He is interested—as he puts it—in the equilateral hyperbola having the lines ox and oy for asymptotes and with the areas bounded below by ox, left and right by verticals, and above by the hyperbola. If the distances of the left and right vertical from oy are, respectively, a and b—therefore, $a < b$—we shall denote this area by $J_{a, b}$. The principle of the reasoning will appear by making some observations on $J_{1, 2}$ and $J_{2, 4}$ (Fig. 36).

Let us divide $[1, 2]$—the interval with two ends 1 and 2—into, say, six equal parts, and $[2, 4]$ also into six equal parts. The latter will be twice as wide as the former, since $[2, 4]$ is twice as wide as $[1, 2]$. We now consider the figures consisting of the partly dotted rectangles belonging to the two intervals. Let their areas be $T_{1,2}$ and $T_{2,4}$, respectively. We now compare rectangle for rectangle. The one farthest to the left in $T_{1,2}$ is half as wide but twice as high as the one farthest to the left in $T_{2,4}$; hence they have equal areas. The second one from the left in $T_{1,2}$ is likewise half as wide as the second one in $T_{2,4}$. The altitude of the former is $1/1\frac{1}{6}$, that of the latter, $1/2\frac{1}{3}$—that is, again, twice as high. And this continues to be true for all the six rectangles. Hence the six rectangles composing $T_{1,2}$ are equal in area to the six composing $T_{2,4}$, that is, $T_{1,2} = T_{2,4}$. The same is true if we divide each of the intervals into n rather than six parts. And, since with increasing n, $T_{1,2}$ approaches $J_{1,2}$, and $T_{2,4}$ approaches $J_{2,4}$, we obtain

$$J_{1,2} = J_{2,4} .$$

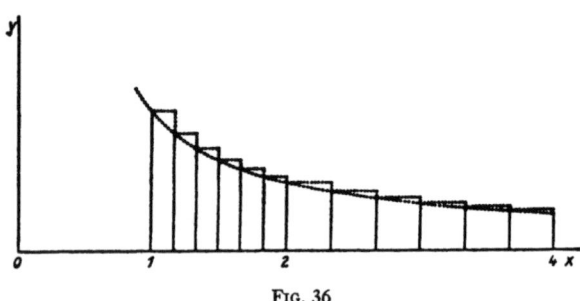

FIG. 36

In exactly the same way we can show $J_{1,2} = J_{3,6}$, in which case the rectangular strips in the one are three times as wide, but only one-third as high; and, more generally,

$$J_{a,b} = J_{ta,tb} .$$

Perhaps we ought to be a little more careful here. In our reasoning above t was an integer; first we had $t = 2$ and then $t = 3$. Does the formula then hold only for integers or for all numbers t? That is easy to answer. First, let t be a fraction, $t = p/q$; then $J_{ta,tb}$ would be

$$J_{(p/q)a,\ (p/q)b} ,$$

which, in turn is equal to

$$J_{p(a/q),\ p(b/q)} ,$$

since $(p/q)a$ is nothing else than $p(a/q)$. But since, for an integer t it had been shown that $J_{a,b} = J_{ta,tb}$, we can now say

$$J_{a/q,\ b/q} = J_{p(a/q),\ p(b/q)} ,$$

which only means that the original interval $[a, b]$ is now $[a/q, b/q]$. Moreover, using once more the fact that our statement had been proved for whole numbers t, we find

$$J_{a/q,\ b/q} = J_{q(a/q),\ q(b/q)} = J_{a,b} ,$$

and hence

$$J_{(p/q)a, \, (p/q)b} = J_{a, \, b} \, .$$

With irrational values t, finally, we would deal by approximating them indefinitely through sequences of rational values. We have thus obtained the *theorem for the segment of the hyperbola:*

$$J_{a, \, b} = J_{ta, \, tb} \, .^*$$

One immediate consequence of this theorem is

$$J_{a, \, b} = J_{1, \, b/a} \, ,$$

which means that all such areas are equal to certain ones, $J_{1, \, x}$, starting at 1.

Now, with x and y both greater than 1, xy is greater than x. The area under the curve from 1 to xy, $J_{1, \, xy}$, however, is equal to the sum of the two parts from 1 to x, $J_{1, \, x}$, and from x to xy, $J_{x, \, xy}$; that is, $J_{1, \, xy} = J_{1, \, x} + J_{x, \, xy}$. But, according to the above theorem, $J_{x, \, xy} = J_{1, \, y}$. This gives finally

$$J_{1, \, xy} = J_{1, \, x} + J_{1, \, y} \, ,$$

which shows that the area under the equilateral hyperbola has the basic property of the logarithm. For how had Napier defined logarithms? He was trying to construct a table which would pair every number x with one—called its logarithm, log x—such that log $xy = $ log $x + $ log y for all values x and y. For, if one has such a table, every multiplication can be reduced to an addition, every division to a subtraction, every kth root to a division by k, etc. Napier (1614) made a study of the principle of such tabulations and found that all the possible ones can be readily derived from a particular one, which he constructed. This, then, was the discovery of Gregorius, that the area $J_{1, \, x}$ under the equilateral hyperbola is exactly this logarithm of Napier. Only the proof was as yet missing that it was the "natural" logarithm to the base e. But that we can discuss only after we are better acquainted with Napier's ideas.

* [The care about t is not needed. For suppose t is a positive number and construct rectangles over the intervals $[a, \, b]$ and $[ta, \, tb]$, each being divided into n equal parts. Let $T_{a, \, b}$ and $T_{ta, \, tb}$ be their areas. Pick out a rectangle in $T_{a, \, b}$ and the corresponding one in $T_{ta, \, tb}$. The first has width $(b - a)/n$; the second, width

$$\frac{tb - ta}{n} = \frac{t(b-a)}{n} = t \, \frac{b-a}{n} \, .$$

If the height of the first is

$$\frac{1}{a + k(b-a)/n} \, ,$$

the height of the second is

$$\frac{1}{ta + k[t(b-a)/n]} = \frac{1}{t} \frac{1}{a + k(b-a)/n} \, .$$

Consequently, the two rectangles have the same area, and it follows that $T_{a, \, b} = T_{ta, \, tb}$, and hence $J_{a, \, b} = J_{ta, \, tb}$.—EDITOR.]

13. AREA AND DEFINITE INTEGRAL

Almost imperceptibly the course of these discoveries had taken a new turn. Beginning with the measurement of areas in general, it had wound up with the study of very special sorts of areas—those, namely, whose boundaries below and on left and on right were box-shaped and which were bounded by a curve only above (Fig. 37). This fact is expressed more systematically in two general lemmas of Cavalieri, the so-called *principles of Cavalieri*. There are two assertions:

1. *Let the box-shaped figure (Fig. 38) bounded by the curve* f *have an area* F *and that bounded by curve* g *the area* G. *Now let a curve* h *be so constructed*

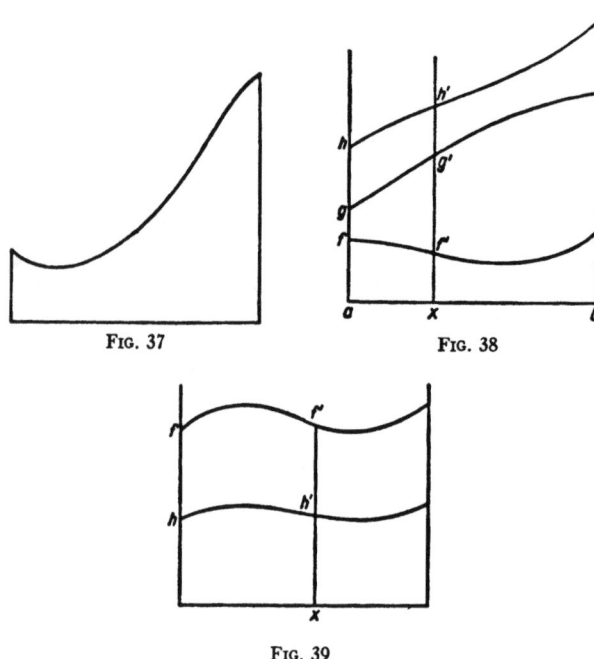

Fig. 37   Fig. 38

Fig. 39

that xh′ = xf′ + xg′ *for every vertical between* a *and* b. *Then the area* H *bounded by the curve* h *is* H = F + G.

2. *If the box-shaped area bounded by* f *is* F, *and if (Fig. 39)* h *is so constructed that, on every vertical between* a *and* b, xh′ = ρxf′, *then* H = ρF. *(In Fig. 39, ρ has been chosen as* ½.)

These two principles have become famous because of the debate over the proof which Cavalieri gave for them. But what was somewhat overlooked in this debate was the important fact that with these principles the turn toward the box-shaped areas had been accomplished, that the *function concept* had entered the scene, a concept whose development is parallel with that of the modern number concept (infinite decimal fraction) and to which it is closely related. In fact, these box-shaped figures are nothing else but the graphical representations of a function f in an interval $a \leq x \leq b$. The curve h of the first principle is the graph of $y = h(x)$, where $h(x) = f(x) + g(x)$; that of the second principle, $h(x) = \rho f(x)$. Nevertheless, it took several more decades before this function concept was algebraically sufficiently developed (Chuquet, Viete, Descartes) for Leibniz to introduce the still customary symbol for the box-shaped area,

$$F = \int_a^b f(x)\, dx.$$

Using this symbol, Cavalieri's two principles take on the form

1. $$\int_a^b [f(x) + g(x)]\, dx = \int_a^b f(x)\, dx + \int_a^b g(x)\, dx$$

2. $$\int_a^b [\rho f(x)]\, dx \qquad = \rho \int_a^b f(x)\, dx \qquad\qquad (\rho > 0).$$

In these symbols it appears clearly how useful a tool Cavalieri had created and what a wealth of problems could be handled by its means, for he knew all the following areas:

$$\int_a^b 1\, dx = b - a, \qquad \int_a^b x\, dx = \frac{b^2 - a^2}{2}, \qquad \int_a^b x^2 dx = \frac{b^3 - a^3}{3}, \ \ldots .$$

The second principle, therefore, gave him at once

$$\int_a^b c_0\, dx = c_0(b - a), \qquad \int_a^b c_1 x\, dx = c_1 \frac{b^2 - a^2}{2},$$

$$\int_a^b c_2 x^2 dx = c_2 \frac{b^3 - a^3}{3}, \ \ldots ,$$

and, if we extend the first principle from two to several summands, we obtain

$$\int_a^b (c_0 + c_1 x + c_2 x^2 + \ldots c_n x^n)\, dx = c_0(b - a) + \frac{c_1}{2}(b^2 - a^2)$$

$$+ \ldots \frac{c_n}{n+1}(b^{n+1} - a^{n+1}) = \left(c_0 b + \frac{c_1 b^2}{2} + \frac{c_2 b^3}{3} + \ldots + \frac{c_n b^{n+1}}{n+1}\right)$$

$$- \left(c_0 a + \frac{c_1 a^2}{2} + \frac{c_2 a^3}{3} + \ldots \frac{c_n a^{n+1}}{n+1}\right).$$

Hence, if we put

$$f(x) = c_0 + c_1 x + c_2 x^2 + \ldots + c_n x^n ,$$

$$\varphi(x) = c_0 x + \frac{c_1 x^2}{2} + \frac{c_2 x^3}{3} + \ldots \frac{c_n x^{n+1}}{n+1},$$

we get

$$\int_a^b f(x)\,dx = \varphi(b) - \varphi(a),$$

where $f(x)$ is an arbitrary polynomial with positive coefficients. Such was the far-reaching result that had grown out of Archimedes' squaring of the parabola. And yet, on glancing back over the whole development, it appears that the decisive step had been exactly the one taken by Archimedes—his discovery, namely, that the squaring of the parabola leads to such a simple result.

14. NON-RIGOROUS INFINITESIMAL METHODS

We now turn to the reasoning used by Cavalieri in establishing his two principles. He says that F is the sum of all the verticals f from a to b of which F consists—therefore, an infinite sum. Hence, if every vertical in H is ρ times the

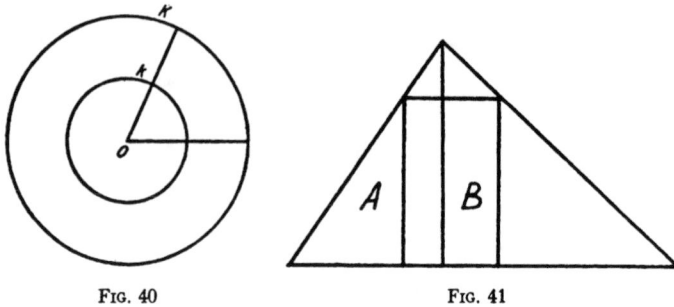

Fig. 40   Fig. 41

corresponding one in F, the sum of all the verticals of H; that is, H itself, is ρ times the sum of the verticals in F, wherefore $H = \rho F$. By the same reasoning he establishes the first principle.

How dangerous this kind of reasoning can be we shall show by two examples. We draw two concentric circles (Fig. 40), K, the larger one, having a radius twice that of the smaller one, L. Now say that the area of K is the sum of all its radii, and the same for L. Hence the sum of all radii of K must be twice that of the radii of L, which gives $K = 2L$. In fact, however, $K = 4L$.

For another example we use a triangle (Fig. 41) whose height divides it into two [unequal] right triangles, A and B. A is the sum of its verticals, of which one is shown, and similarly for B. Such verticals as shown in the figure are equal to each other.* Thus to every vertical in A there belongs one definite vertical in B to which it is equal. Hence, with all verticals being pair-wise equal in A and B, their sums are equal, and $A = B$. Obvious nonsense! The false result in

---

* [Given a vertical in A, we construct a vertical in B by translating the former parallel to itself in the direction determined by the base of the large triangle, until it meets the hypotenuse of B.—Editor.]

the case of the circles, therefore, did not result merely from the fact that the radii were non-parallel.

The ingenious method of Cavalieri with his "indivisibles," as he called the lines, was thus highly questionable; nor did he succeed in refining it so as to exclude the two fallacies just mentioned. He had only his "sound instinct" to rely on, while his contemporaries, who had been brought up on rigorous Euclidean mathematics, widely debated the validity of his thesis. We should, however, gain a wrong impression of the age if we believed that Cavalieri had, stood alone with his tendencies. Guldin, with his rule for the volume of the torus; the Frenchman Pascal; the great Kepler—all pursued similar paths. And most of all Galileo, greatest among all of them, was foremost in this revolutionary movement: away with Aristotle, off with the fetters of rigorous Greek procedures!

In fact, these "fetters" were an optical illusion produced by scholasticism. Scholasticism had handed down a petrified image of Greek scientific thought; it had made Aristotle into an orthodoxy and treated Euclid's text as *"The* Mathematics." This tendency had set in already in ancient times, when there was no longer enough genius to deal freely with ideas in the way Plato, Aristotle, Eudoxus, Euclid, or Archimedes had done.

In truth the Greeks used indivisibles, and quarreled over their use, just as the first half of the seventeenth century did. In 1906 the manuscript of a treatise by Archimedes which had been lost until then was found in Constantinople and whose existence had been known only from some ancient quotations. In this work Archimedes does not give any new proofs but only wants to show by what methods of reasoning he had found all his great results; hence the treatise is called ἔφοδος,[27] "the way, the method, the approach." And what is this approach? The method of indivisibles, which he, however, handles with a boldness and instinctive security which leaves Cavalieri far behind. The segment of the parabola he actually found in this way, too, not through the artful proof given above, which was manufactured afterward.

And Archimedes reveals more. He reveals that some theorems (e.g., those concerning the volume of the cone and others) which had first been proved rigorously by Eudoxus had been discovered before, namely, by Democritus. He does not say that Democritus found them by the method of indivisibles. But there exists a fragment of a work by Democritus which points toward his having used indivisibles. To be sure, the matter is not clear. Democritus was an "atomist." But whether he meant only that matter consists of small indivisible parts, or whether he also thought of geometrical (massless) space as being atomistically constituted, so as to consist of a very great but *finite* number of smallest parts—that I have never been able to find out from all the print spent on it. This idea, by the way, would certainly not be identical with the idea of indivisibles, whose number is regarded as *infinite*. An area consists of infinitely many line segments!

At any rate, Greek antiquity debated these questions with great passion.

Democritus was anathema to Plato—exactly why, we do not know—for by never mentioning him he disregarded his very existence. Plato's successors, notably Xenocrates, wrote extensive treatises on indivisibles. Aristotle, too, debated them. But what was handed down to posterity was only that which was not debatable: Euclid and the works of Archimedes exclusive of the ἔφοδος. And this is what produced that optical illusion of Greek mathematics as rigid which dominated the Middle Ages and against which the age of Galileo rebelled with such vehemence.

The seventeenth century, unaware of the antiquity of the issue, adopted the indivisibles. All these tendencies then come to a focus in Leibniz, as in a light-gathering lens, and radiate from him throughout the eighteenth century. His successors—Euler, the Bernoullis, Taylor, and others—then create, with the uninhibited zeal of discovery, the edifice of the new mathematics on the foundation of such non-rigorous, heuristic methods. By 1800 it appears that one cannot go on like this, and now one gradually learns to combine rigorous Greek form with the heuristic fruitfulness of the indivisibles. Eventually this is accomplished; and thus we have today within the reach of any mathematician that perfection of method which, in ancient times, only the towering intellect and genius of Archimedes could achieve.

15. THE CONCEPT OF THE DEFINITE INTEGRAL

If in an examination you ask a student of average ability the question, "What is a definite integral?" he takes his pencil, draws a box-shaped figure, shades it, and says: "This area." When you next ask him, "What is an area?" he is puzzled and says, perhaps, that it is a geometric concept. And when you next insist that, whether geometrical or analytical, every mathematical concept must be precisely defined, he will say: "All right, an area is a definite integral."

This shows the need for defining one of the two concepts clearly and independently of the other. We might do it for the area concept, analyzing what we have so far asserted about areas and what tacit assumptions were at the bottom of such assertions. For several reasons I prefer to define the definite integral. We note in advance that the nature of this definition is not exactly simple.

To define a square number is an easy matter: A square number is a number which results from multiplying a number by itself. That is unambiguous: 16 is a square number, 17 is not, and every number either is or is not a square number. When it comes to square roots, and we limit ourselves to whole numbers, it is quite another matter. $\sqrt{a}$ means the whole number whose square equals a. But is there always such a number? Obviously not! The definition, therefore, means in fact, "If there is a number which . . . , then it is called the square root of a." It is this kind of definition which will have to be used for the definite integral and later for the differential quotient. The definition of convergence had also been of this kind. These kinds of definitions are really not so much definitions as prescriptions for definitions. The prescription can be filled only when one is ready to take the medicine.

I can write down the prescription for the definition of the definite integral very simply, as follows: Let there be an interval $a \leq x \leq b$ in which the function f is defined. I divide the interval first into two, then into four, eight, etc., equal parts, and compute the areas of the lower and upper polygons (Fig. 42)

$$S_n = (x_1 - a)f(a) + (x_2 - x_1)f(x_1) + \ldots + (b - x_{n-1})f(x_{n-1}) \,,$$

$$T_n = (x_1 - a)f(x_1) + (x_2 - x_1)f(x_2) + \ldots + (b - x_{n-1})f(b) \,,$$

for $n = 2, 4, 8, \ldots$. Consider the two sequences $S_2, S_4, S_8, \ldots$, and $T_2, T_4, T_8, \ldots$. If both converge toward one and the same limit, I call this limit the definite integral, and write

$$J = \int_a^b f(x)\,dx = \lim_{n \to \infty} S_n = \lim_{n \to \infty} T_n \,.$$

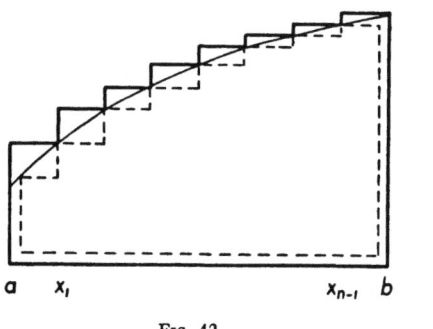

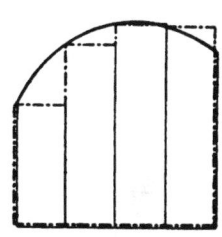

Fɪɢ. 42   Fɪɢ. 43

But the question always is whether they do converge. That is why I said that I would give only a prescription for a definition, not a definition as such. The drawing of the figure could easily deceive us into regarding convergence as readily recognizable. However, Figure 43 shows that the S_n need not be "contained in the area" and that the T_n need not "contain" it. The idea that the S_n always increase, and the T_n always decrease with increasing n, appears at once as erroneous. The reason for this is also at once clear: In Figure 43 the curve changes from increasing to decreasing; in Figure 42 it does not. This leads us to the following statement: If the curve always increases from a to b, then (1) $f(x_{i-1}) \leq f(x_i)$, and hence every one of the rectangles of which S_n consists is less than the corresponding rectangle belonging to T_n; (2) it appears that under such circumstances $S_2 \leq S_4 \leq S_8 \ldots$, and $T_2 \geq T_4 \geq T_8 \ldots$. Referring back to Theorem I on convergent sequences, we find that three of the conditions for convergence are satisfied. The question is only whether the fourth one is also satisfied.

$$\lim_{n \to \infty} (T_n - S_n) = 0 \,.$$

This is at once seen, for

$$T_n - S_n = (x_1 - a)[f(x_1) - f(a)] + \ldots + (b - x_{n-1})[f(b) - f(x_{n-1})] \,.$$

Since all intervals $x_i - x_{i-1}$ are each equal to the nth part of ab, that is, each equal to $(b - a)/n$, we obtain

$$T_n - S_n = \frac{b-a}{n}\Big\{[f(x_1) - f(a)] + [f(x_2) - f(x_1)]$$
$$+ \ldots + [f(b) - f(x_{n-1})]\Big\}$$
$$= \frac{b-a}{n}[f(b) - f(a)] \,.$$

This same result can also be inferred from Figure 44, in which $T_n - S_n$ is represented by the totality of the shaded rectangles all of which have the same width and whose heights add up to the total increase of the function $f(x)$ from

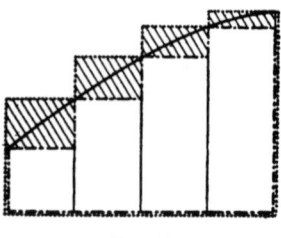

Fig. 44

a to b. If now n increases, a, b, $f(a)$, and $f(b)$ all remain fixed; hence $T_n - S_n$ does indeed approach zero. Theorem I (chap. i), therefore, applies, and—under the given conditions—we do, indeed, obtain a definition for the definite integral.

THEOREM 1. *If* f(x) *is monotonic in the interval* ab, *that is, if* u $\leq$ v *always implies* f(u) $\leq$ f(v), *then*

$$\lim_{n \to \infty} S_n = \lim_{n \to \infty} T_n = \int_a^b f(x)\,dx \,,$$

and hence this integral exists.

We shall next have to prove some theorems concerning the definite integral. In doing so, we would, however, encounter a difficulty arising from the fact that there was something arbitrarily restrictive in the way in which we formulated our definition. We divided the interval only into equal parts, 2, 4, 8, ... In the preceding section, however, some of the best results were obtained when the interval was divided into unequal parts, even into infinitely many such. It seems advisable, therefore, to recast the definition of the definite integral so as to allow for unequal subdivisions, at least into a finite number of subintervals.

Let us denote any subdivision of ab (Fig. 45) into a finite number of parts by a capital Latin letter. The division into n equal parts we call E_n. If A is any other subdivision of ab, then we write

$$S_A = (x_1 - a)f(a) + \ldots + (b - x_{n-1})f(x_{n-1}) \,,$$

$$T_A = (x_1 - a)f(x_1) + \ldots + (b - x_{n-1})f(b) \,.$$

Let d_A be the width of the widest subinterval in A. The sums which we considered above, corresponding to subdivisions into equal subintervals, were—in our present notation—$S_{E_2}, S_{E_4} \ldots, T_{E_2}, T_{E_4} \ldots$, and d_{E_n} was $(b - a)/n$.

Instead of the successive subdivisions $E_2, E_4, E_8, \ldots$, we now consider some other sequence of successive subdivisions, $A_1, A_2, A_3, \ldots$, and assume again that

$$\lim_{n \to \infty} d_{A_n} = 0 \;;$$

that is, that under continued subdividing, the maximum width of the subintervals becomes ever smaller. Under these assumptions we now investigate whether the sequences $S_{A_1}, S_{A_2}, \ldots$, and $T_{A_1}, T_{A_2}, \ldots$, converge to $\int_a^b f(x)dx$, too.

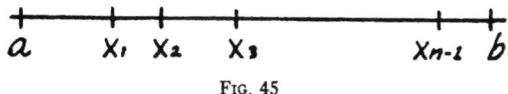

FIG. 45

To begin with, we restrict ourselves again to monotonic functions. There is a difficulty here: We do not wish to assume about $A_1, A_2, \ldots$, that A_{n+1} contains all points of division occurring in A_n. For the $E_2, E_4, E_8, \ldots$, this was the case, since E_8 results from E_4 by adding to the three points of division four new ones. But, even so, if we were to consider all the partitions $E_1, E_2, E_3, \ldots$, this would not any longer be true. And we are interested in a definition which is free from such a restrictive assumption.

It is not difficult to achieve this. If f is monotonic, two things are obviously true, as above:

1. $S_A \leq T_A$ for any subdivision A.

2. If A is contained in the subdivision C, that is, if all division points of A are also division points of C, then it is true that

$$S_A \leq S_C \text{ and } T_A \geq T_C \,.$$

But we must be sure that for any two subdivisions, A and B, $S_A \leq T_B$ will always hold. Geometrically, this seems obvious enough, since S_A lies completely "inside" the "figure," while T_B "contains" it. But what we want is a derivation free from representational elements. Otherwise, we might indeed simply have said that an integral is an area—period! We can tackle it, however, in the following way: Let C be a subdivision which consists of all the division points of A and B together; then both A and B are contained in C. Hence, because of **2,** we have

$$S_A \leq S_C, \quad T_B \geq T_C,$$

and, because of **1,**

$$S_C \leq T_C;$$

hence

$$S_A \leq S_C \leq T_C \leq T_B,$$

which is what we wanted to prove.

From here on everything is easy. For again, as before, one has for every subdivision A

$$T_A - S_A = (x_1 - a)[f(x_1) - f(a)] + \ldots + (b - x_{n-1})[f(b) - f(x_{n-1})].$$

Now, of course, the numbers $x_1 - a, \ldots, b - x_{n-1}$ are not all equal to one another and hence not equal to $(b - a)/n$; but they are all at most equal to d_A, in accordance with the definition of this symbol, and hence

$$T_A - S_A \leq d_A[f(b) - f(a)].$$

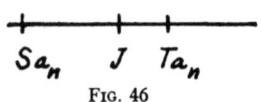

$$Sa_n \quad J \quad Ta_n$$

FIG. 46

This holds for $A = A_1, A_2, \ldots$. Since $f(b) - f(a)$ remains fixed under any of these schemes of subdivision, while d_A, according to hypothesis, approaches zero, $T_A - S_A$ goes to zero. And now we can argue as follows: For all A_n we have $S_{A_n} \leq T_{B_1}, T_{B_4}, \ldots$, according to our above result. Hence also $S_{A_n} \leq J = {}^b\int_a f(x)dx$; similarly, for all A_n, we have $T_{A_n} \geq S_{B_1}, S_{B_4}, \ldots$, and hence also $T_{A_n} \geq J = {}^b\int_a f(x)dx$ (Fig. 46). But $T_{A_n} - S_{A_n}$ decreases indefinitely with increasing n; consequently, a fortiori, $J - S_{A_n}$ and $T_{A_n} - J$ approach zero with increasing n; that is,

$$\lim_{n \to \infty} S_{A_n} = \lim_{n \to \infty} T_{A_n} = \int_a^b f(x)\,dx.$$

THEOREM 2. *If* f(x) *is monotonic for* a $\leq$ x $\leq$ b, *and if* $A_1, A_2, \ldots$, *is some one sequence of subdivisions of the interval* ab, *whose maximal subinterval* d_{A_n} *converges to zero, then*

$$\lim_{n \to \infty} S_{A_n} = \lim_{n \to \infty} T_{A_n} = \int_a^b f(x)\,dx.$$

We are now sufficiently prepared to do away with the restrictive assumption that $f(x)$ is monotonic—not do away with it altogether but relax it so as to give us full freedom for all our purposes in this course. Let $f(x)$ be not altogether monotonic in ab but (Fig. 47) monotonically increasing from a to c and monotonically decreasing from c to b. (For functions which *decrease* monotonically in an interval, everything remains true which was found for monotonically increasing ones.) If now we consider, for example, the subdivision E_8 of the interval ab, it produces subdivisions in ac and cb, too; in general, however, these subintervals will not be equal unless, by chance, c happens to be one of the division points of E_8. Now it becomes clear why we needed to know how to handle unequal subintervals. Let A_2, A_4, . . . , be the subdivisions produced by E_2, E_4, . . . , in ac, and B_2, B_4, . . . , those produced in cb. Then

$$S_{E_i} = S_{A_i} + S_{B_i} ,$$

and, in the limit

$$\int_a^b f(x)\,dx = \int_a^c f(x)\,dx + \int_c^b f(x)\,dx ;$$

that is, according to Theorem 2. S_{A_2}, S_{A_4}, . . . , converge to $^c\!\int_a f(x)dx$ and S_{B_2}, S_{B_4}, . . . , to $^b\!\int_c f(x)dx$. We can reason similarly if, instead of S_{E_2}, S_{E_4}, . . . , we

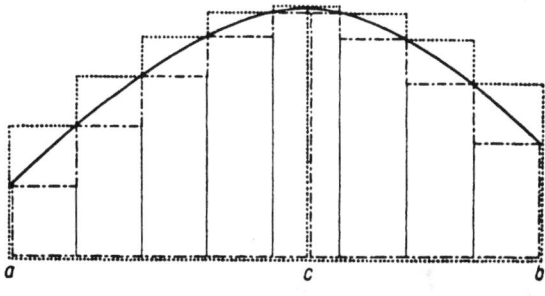

Fig. 47

consider any other sequence of subdivisions of ab. Finally, everything will remain valid if $f(x)$ changes in ab from rising to falling or vice versa not just once but several times, provided only that it happens a *finite* number of times.

THEOREM 3. *If a function* f(x) *is "sectionwise monotonic" in an interval* ab, *that is, if we can divide the interval into a finite number of sections such that* f(x) *is monotonically increasing or monotonically decreasing in each of these, then the assertions of statements 1. and 2. hold for this function.*

There are situations in higher mathematics where to be sectionwise monotonic is not enough—where a sharper formulation of the above statements is needed. This, however, requires more mathematical maturity than can be assumed here. Whoever studies those topics of higher mathematics needs, of course, to possess this maturity. But I regard it as a mistake when textbooks, disregarding the need for balancing methodological difficulties, bring in here a bulky theorem which can be neither used nor digested at this stage.

In concluding this section, I want to come back once more to the distinction between area and definite integral and to show that they are decidedly not the same thing. Let us consider $\int_{-1}^{1} x\,dx$. We divide the interval into $2n$ equal parts (Fig. 48). In the case of 12, for example,

$$S_{E_{12}} = \frac{1}{6}\left(\frac{-6}{6}\right) + \frac{1}{6}\left(\frac{-5}{6}\right) + \cdots + \frac{1}{6}\left(-\frac{1}{6}\right)$$
$$+ \frac{1}{6}(0) + \frac{1}{6}\left(\frac{1}{6}\right) + \cdots \frac{1}{6}\left(\frac{5}{6}\right),$$

$$T_{E_{12}} = \frac{1}{6}\left(\frac{-5}{6}\right) + \frac{1}{6}\left(\frac{-4}{6}\right) + \cdots \frac{1}{6}(0) + \frac{1}{6}\left(\frac{1}{6}\right) + \frac{1}{6}\left(\frac{2}{6}\right) + \cdots + \frac{1}{6}\left(\frac{6}{6}\right);$$

from which

$$S_{E_{12}} = \frac{1}{6}\left(\frac{-6}{6}\right) = -\frac{1}{6}, \qquad T_{E_{12}} = \frac{1}{6}\left(\frac{6}{6}\right) = \frac{1}{6}.$$

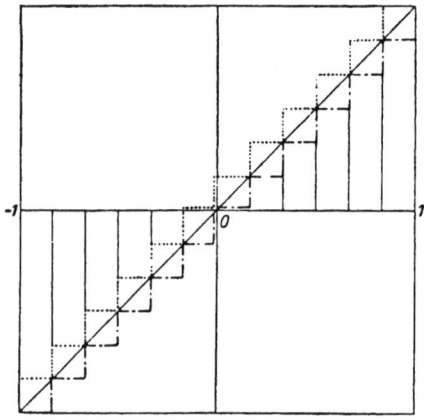

FIG. 48

More generally,

$$S_{E_{2n}} = \frac{-1}{n}, \quad T_{E_{2n}} = \frac{1}{n}: \quad \lim_{n\to\infty} S_{E_{2n}} = 0, \quad \lim_{n\to\infty} T_{E_{2n}} = 0; \quad J = \int_{-1}^{1} x\,dx = 0.$$

How can an area be zero? We see at once where the trouble lies. According to our definition of the integral, strips which lie below the x-axis are counted as negative. This means that

$$\int_{0}^{1} x\,dx = \tfrac{1}{2}, \qquad \int_{-1}^{0} x\,dx = -\tfrac{1}{2};$$

their sum is zero.

THEOREM 4. $\int_{a}^{b} f(x)\,dx$ *is the area of those parts of the box-shaped figure which lie above the x-axis diminished by the area of those which lie below the x-axis.*

16. SOME THEOREMS ON DEFINITE INTEGRALS

THEOREM 1. *If the definite integrals* $^b\!\int_a f(x)dx$ *and* $^b\!\int_a g(x)dx$ *exist, then the integral* $^b\!\int_a [f(x) + g(x)]dx$ *also exists and is equal to* $^b\!\int_a f(x)dx + ^b\!\int_a g(x)dx$.

Proof. We say that $^b\!\int_a f(x)dx$ "exists" if

$$\lim_{n\to\infty} S_{E_{2n}} \quad \text{and} \quad \lim_{n\to\infty} T_{E_{2n}}$$

exist and are equal to each other. This, by hypothesis, is the case for both f and g. However, for each subdivision A, $S_A(f + g) = S_A(f) + S_A(g)$. This, according to Theorem 1 on convergent sequences (Sec. 9), proves our statement.

In a similar manner we prove also the following:

THEOREM 2. *If* $^b\!\int_a f(x)dx$ *exists, then* $^b\!\int_a [\rho f(x)]dx$ *also exists and is equal to* $\rho^b\!\int_a f(x)dx$.

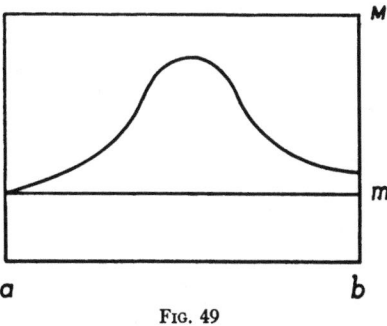

FIG. 49

Applying these theorems repeatedly, we obtain:

THEOREM 3. *If* $^b\!\int_a f_1(x)dx, \ldots, ^b\!\int_a f_n(x)dx$ *exist, then* $^b\!\int_a [\rho_1 f_1(x) + \ldots + \rho_n f_n(x)]dx$ *exists and is equal to* $\rho_1 ^b\!\int_a f(x)dx + \ldots + \rho_n ^b\!\int_a f_n(x)dx$.

THEOREM 4. *If for every x in the interval* ab *it is the case that* $f(x) \leq g(x)$, *then* $^b\!\int_a f(x)dx \leq ^b\!\int_a g(x)dx$, *provided both integrals exist.*

This statement follows directly from Theorem 5 on convergent sequences (Sec. 9). For from the hypothesis we obtain, for each subdivision A, $S_A(f) \leq S_A(g)$, and the theorem mentioned permits us to pass from this inequality to that of the limits, for $A = E_2, E_4, \ldots$.

THEOREM 5. *If in the whole interval* ab *it is true that* $m \leq f(x) \leq M$, *then* $m(b - a) \leq ^b\!\int_a f(x)dx \leq M(b - a)$. *(See Fig. 49.)*

Proof. If in Theorem 4 above we replace $g(x)$ by the constant function M, then $^b\!\int_a g(x)dx = ^b\!\int_a M dx$. According to Theorem 2, $^b\!\int_a M dx = M ^b\!\int_a dx$; finally, $^b\!\int_a M dx = M(b - a)$. This proves the right side of the statement; the left side is obtained by substituting, in Theorem 4, m for $f(x)$ and $f(x)$ for $g(x)$.

THEOREM 6. *If in the whole interval* $|f(x)| \leq \mu$, *then* $|{}^b\!\int_a f(x)dx| \leq \mu(b - a)$.

Since $|f(x)| \leq \mu$ means $-\mu \leq f(x) \leq +\mu$, according to Theorem 5, $-\mu(b - a) \leq {}^b\!\int_a f(x)dx \leq \mu(b - a)$; this pair of inequalities can again be combined in the form of the statement above.

THEOREM 7. *If* a $<$ c $<$ b, *and if* f(x) *is sectionwise monotonic in the whole interval* ab, *then* ${}^b\!\int_a f(x)dx = {}^c\!\int_a f(x)dx + {}^b\!\int_c f(x)dx$.

For even if the subdivision of *ab* into equal parts should bring about unequal subintervals in *ac* and *cb*, Theorem 2 of the preceding section permits us to arrive at the statement in question.

So far the definition of the definite integral always assumed $a < b$. We now go further:

DEFINITION. ${}^a\!\int_a f(x)dx = 0$; ${}^b\!\int_a f(x)dx = -{}^a\!\int_b f(x)dx$, a $>$ b.

As we can readily see, on the basis of this definition, Theorem 7 remains valid for any three numbers *a*, *b*, *c*, regardless of their relative position.

COROLLARY. *Theorem 7 holds for any three numbers* a, b, c.

17. QUESTIONS OF PRINCIPLE

We insert at this point two digressions which throw more light on the concepts we are developing but which do not produce any new facts. The one digression is abstract in nature. We proceeded very cautiously in "prescribing" the definite integral, but was all this caution really necessary? Why not simply prove, once for all, that ${}^b\!\int_a f(x)dx$ exists for *any* sequence of subdivisions of the interval?

But what after all, is a function? (We traced the idea historically and are going to treat it further in the next chapter.) So far we have not even given a precise formulation of the notion of a function. We might say that "in the interval $a \leq x \leq b$ a function f is defined" means that to every value x in this interval belongs a definite number $f(x)$. That is clear and simple and the widest possible definition of a function. Under it fall a multitude of relations which we may hardly have had in mind at first.

By way of example, we define the following function in the interval $0 \leq x \leq 1$: for every irrational number x in the interval $f(x) = 0$; for rational numbers $x = p/q$ likewise $f(x) = 0$, except that when p/q, in lowest terms, is such that q is a square number, then $f(x) = 1$. That is a perfectly well-defined function:

$$f(0) = 1 , \quad f(1) = 1 , \quad f(\tfrac{1}{2}) = 0 , \quad f(\tfrac{1}{4}) = 1 , \quad f(\tfrac{3}{4}) = 1 ,$$

$$f(\tfrac{1}{8}) = 0 , \quad f(\tfrac{3}{8}) = 0 , \quad f(\tfrac{5}{8}) = 0 , \quad f(\tfrac{7}{8}) = 0 ,$$

$$f(\tfrac{1}{16}) = 1 , \quad f(\tfrac{3}{16}) = 1 , \cdots , \quad f(\tfrac{15}{16}) = 1 , \text{ etc.}$$

Now let us form the sequence of sums $S_2, S_4, \ldots$ (Fig. 50). We find:

$$S_2 = (2-1)\tfrac{1}{2}$$

$$S_{2^2} = (4-1)\tfrac{1}{4}, \qquad S_{2^3} = (4-1)\tfrac{1}{8}$$

$$S_{2^4} = (12-1)\tfrac{1}{16}, \qquad S_{2^5} = (12-1)\tfrac{1}{32}$$

$$S_{2^6} = (44-1)\tfrac{1}{64}, \qquad S_{2^7} = (44-1)\tfrac{1}{128}$$

$$S_{2^8} = (172-1)\tfrac{1}{256}, \qquad \ldots$$

Computing gives for the numerators:

$$4-1=3=1+2$$

$$12-1=11=1+2+8 \qquad\qquad =1+2(1+4) \qquad\qquad =1+2\,\frac{4^2-1}{4-1}$$

$$44-1=43=1+2+8+32 \qquad =1+2(1+4+16) \qquad =1+2\,\frac{4^3-1}{4-1}$$

$$172-1=171=1+2+8+32+128=1+2(1+4+16+64)=1+2\,\frac{4^4-1}{4-1}$$

$\ldots$

$$0 \qquad\qquad\qquad\qquad\qquad\qquad 1$$

Fig. 50

Substituting, we then have:

$$S_{2^2}=\frac{1+2}{4}, \qquad\qquad S_{2^3}=\tfrac{1}{2}S_{2^2}$$

$$S_{2^4}=\frac{1+\tfrac{2}{3}(4^2-1)}{4^2}, \qquad S_{2^5}=\tfrac{1}{2}S_{2^4}$$

$$S_{2^6}=\frac{1+\tfrac{2}{3}(4^3-1)}{4^3}, \qquad S_{2^7}=\tfrac{1}{2}S_{2^6}$$

$$S_{2^8}=\frac{1+\tfrac{2}{3}(4^4-1)}{4^4}, \qquad \ldots.$$

It appears now that $S_{2^2}, S_{2^4}, S_{2^6}, \ldots$, converge to $\tfrac{2}{3}$, while $S_{2^3}, S_{2^5}, S_{2^7}, \ldots$, converge to $\tfrac{1}{3}$. Thus the whole sequence $S_2, S_4, S_8 \ldots$ oscillates perpetually between $\tfrac{2}{3}$ and $\tfrac{1}{3}$, and hence there is no convergence of the sequence as a whole.

Sequences resulting from other methods of subdivision do not lead any further. If, for the subdivision, only irrational points were chosen S_{A_n} would converge to zero; for a certain other choice, 1 would be the limit. In fact, by carefully choosing the sequences, any number between 0 and 1 could be obtained as a limit, and so any kind of oscillation.

As another, very different, example we take $f(x) = 1/x$ for $0 < x \leq 1$, and $f(0) = 0$. This defines a "function" for $0 \leq x \leq 1$, in that to every x in the interval belongs a number $f(x)$ (Fig. 51).

We form the sequence $S_2, S_4, \ldots$, and find

$$S_2 = \tfrac{1}{2}\cdot 0 + \tfrac{1}{2}\cdot 2 \qquad\qquad\qquad\qquad = 1$$

$$S_4 = \tfrac{1}{4}\cdot 0 + \tfrac{1}{4}\cdot\tfrac{4}{1} + \tfrac{1}{4}\cdot\tfrac{4}{2} + \tfrac{1}{4}\cdot\tfrac{4}{3} \qquad = 1 + \tfrac{1}{2} + \tfrac{1}{3}$$

$$S_8 = \tfrac{1}{8}\cdot 0 + \tfrac{1}{8}\cdot\tfrac{8}{1} + \tfrac{1}{8}\cdot\tfrac{8}{2} + \ldots + \tfrac{1}{8}\cdot\tfrac{8}{7} \qquad = 1 + \tfrac{1}{2} + \tfrac{1}{3} + \ldots + \tfrac{1}{7}$$

$$S_{16} = \tfrac{1}{16}\cdot 0 + \tfrac{1}{16}\cdot\tfrac{16}{1} + \tfrac{1}{16}\cdot\tfrac{16}{2} + \ldots + \tfrac{1}{16}\cdot\tfrac{16}{15} = 1 + \tfrac{1}{2} + \tfrac{1}{3} + \ldots + \tfrac{1}{15}.$$

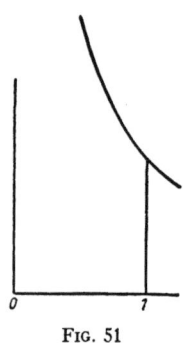

FIG. 51

But

$$1 + \tfrac{1}{2} + \tfrac{1}{3} > 1 + \tfrac{1}{2},$$

$$1 + \tfrac{1}{2} + (\tfrac{1}{3} + \tfrac{1}{4}) + \tfrac{1}{5} + \tfrac{1}{6} + \tfrac{1}{7} > 1 + \tfrac{1}{2} + (\tfrac{1}{4} + \tfrac{1}{4}) = 2,$$

$$1 + \tfrac{1}{2} + (\tfrac{1}{3} + \tfrac{1}{4}) + (\tfrac{1}{5} + \tfrac{1}{6} + \tfrac{1}{7} + \tfrac{1}{8}) + \ldots + \tfrac{1}{15}$$

$$> 1 + \tfrac{1}{2} + (\tfrac{1}{4} + \tfrac{1}{4}) + (\tfrac{1}{8} + \tfrac{1}{8} + \tfrac{1}{8} + \tfrac{1}{8}) = 2\tfrac{1}{2}.$$

The next sum is greater than 3, the next greater than $3\tfrac{1}{2}$, 4 . . . ; that is, the numbers S_{2^n} grow beyond any bound. The sequence does not converge.

The reason for the non-convergence is not that the values of the function in this example are unbounded. In another example of a function with unbounded values the limit does exist: In the interval $0 \leq x \leq 1$,

$$f(0) = 0, \qquad f(1) = 1, \qquad f(x) = (\sqrt{2})^n$$

$$\text{for } \frac{1}{2^{n+1}} \leq x \leq \frac{1}{2^n}, \qquad n = 0, 1, 2, \ldots .$$

Here we obtain:

$$S_{2^n} = \tfrac{1}{2}\left(1 + \frac{1}{\sqrt{2}} + \ldots + \frac{1}{\sqrt{2^{n-1}}}\right), \qquad T_{2n} = S_{2^n} + \frac{1}{2^n}.$$

Both sequences converge to the limit $1/(2 - \sqrt{2})$, even though the function is unbounded. The first example shows, on the other hand, that the sequences may be divergent even when the function is bounded.

Thus the assumption that the integral always exists appears to be a delusion. The question whether, in a given situation, it does or does not exist is a very complex one. Hence our caution all along, and our talking of "prescriptions for definition!"

The other digression is historical in nature. It concerns the relation of the modern concept of the integral and the Greek concept of area. For the difficulties which we have just pointed out must have somehow occurred to the Greeks, keen thinkers that they were, whether they talked of areas or of integrals. (This was indeed the case, and the comparison is very interesting.)

The Greeks began all geometrical definitions with axioms or postulates. They made certain statements concerning points, straight lines, line segments, angles, etc., and based their proofs on these same statements; for example, that two

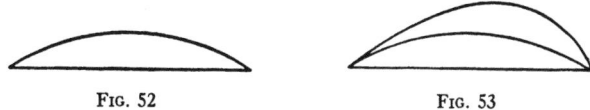

FIG. 52   FIG. 53

points always determine one straight line. That is well known. Less well known are their postulates on area:

1. If one area contains another, the first is greater.

2. If an area consists of two parts, it is equal to the sum of the areas of the parts.

In proving a theorem concerning the area of a circle or some other figure, Euclid did not question whether this area "exists" but talked about "the area" as of something existing as a matter of course. He applied the above two postulates until his statement was proved, and that process proved the "existence" of the area in each particular case. Also a "prescription for a definition"—but certainly different.

Archimedes never questioned Euclid's mathematics in any way but proceeded within its framework. In determining areas, he followed Euclid entirely, but when he undertook to define lengths of arcs and curved surfaces, he did not find the proper models with Euclid and hence was "forced" to proceed in his own way. And with that the story becomes interesting.

Archimedes formulated two postulates for lengths of arcs,[28] just as Euclid did for areas:

1. An arc is longer than the chord joining its end points (Fig. 52).

2. If one arc includes another arc, the first is longer than the second (Fig. 53).

The second postulate, of course, appears to be false even in simple cases: A includes B, but B surely can be made longer than A simply by twisting it sufficiently (Fig. 54). But Archimedes did not formulate the two axioms in such general terms. First, he assumed arcs to be "convex"; he was not interested in other arcs. In contrast to Euclid, he thus introduced a limitation to his procedure just as we did in the case of the definite integral when we required $f(x)$ to be monotonic. Archimedes knew well that without such precaution things might go wrong. Whether that was merely a sound instinct of his or whether he was aware of all the ramifications is not evident.

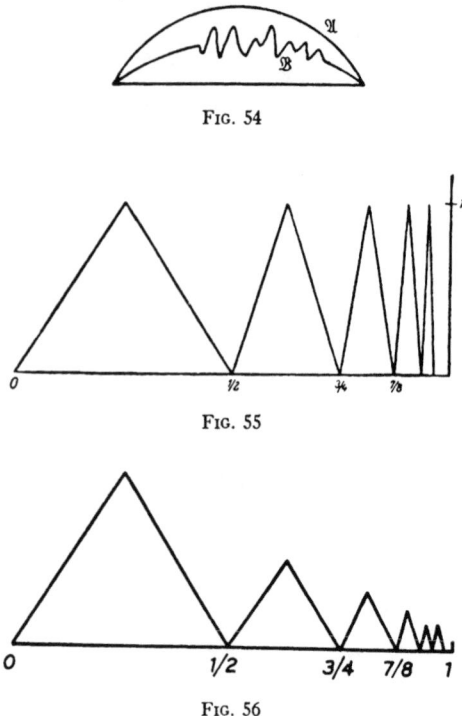

FIG. 54

FIG. 55

FIG. 56

In fact, the "length of an arc" can become meaningless if the curve is not convex (Fig. 55). This polygonal line consists of infinitely many segments, each greater than 1; its "length" is therefore infinite. Even if the heights do not remain constant but are decreasing so that the nth height is $1/n$ (Fig. 56), the arc length would be greater than $1 + \frac{1}{2} + \frac{1}{3} + \ldots + 1/n$, which, with increasing n, increases without limit.

What is a "convex curve" after all? Archimedes gave a marvelously clever definition: "A plane region is convex if it contains, with any two of its points A and B, also the whole-line segment AB between them." (Figure 57 shows, on the right, a region where the condition is not satisfied for two points A and B, and which, therefore, is not convex.) To Archimedes, then, an arc of a curve is convex if it forms, together with the boundary chord which joins its endpoints, a convex region (Fig. 58).

The following theorem is the counterpart of Theorems 1 and 2 of Section 15. Archimedes did not prove this theorem, nor even state it, but he used it implicitly for each particular figure. He could have proved it; he was not interested in expressing himself in such a generalized manner.

Let AB be a convex arc of a curve (Fig. 59). We assert that this curve has a definite length. First, the curve must lie completely on one side of the line through A, B. For if it had points C, D on both sides, the whole chord CB would belong to the region, and AB would not be the convex boundary chord (Fig. 60). Second, we must assume that the curve should not be infinitely extended, although this need not contradict the condition of convexity (Fig. 61). We thus

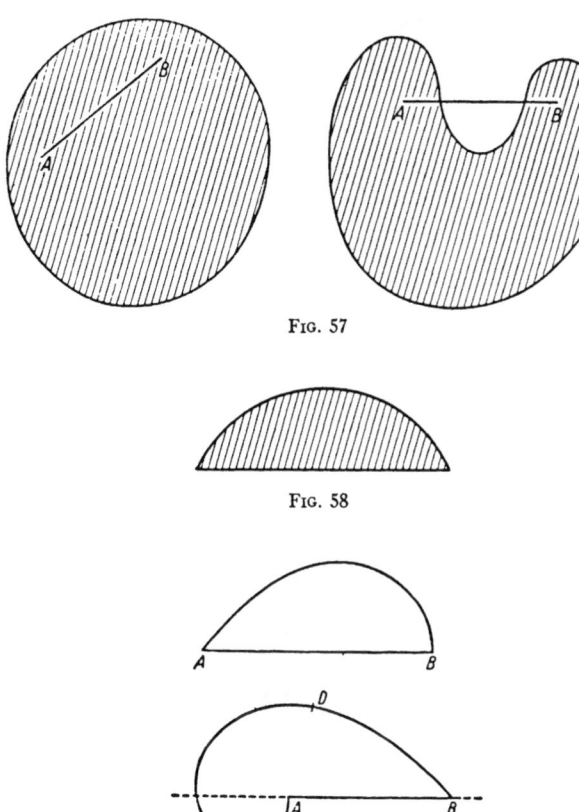

Fig. 57

Fig. 58

Figs. 59 and 60

assume that we can describe a circle around the midpoint M of AB which completely contains the curve (Fig. 62). The shaded convex arc from A to B thus contains the given arc C and is longer than C.

Now we consider a polygon P_n inscribed in C. Its perimeter is equal to the sum of the segments of which it consists. If L is the length of the polygon which includes the shaded arc, we can prove from elementary geometry that L is longer than P_n, as was postulated in Archimedes' second axiom. (This is simply

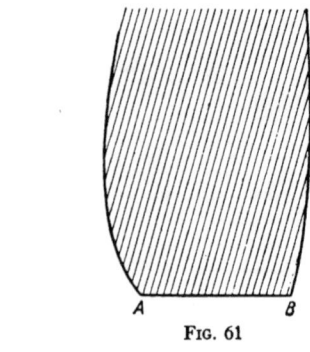

FIG. 61

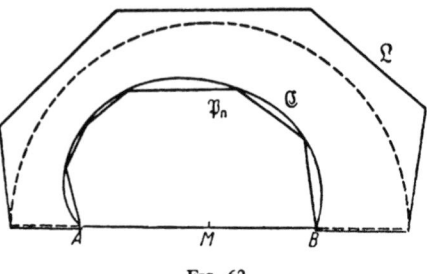

FIG. 62

proved by the theorem that in any triangle one side is shorter than the sum of the other two.)

All P_n, no matter how large n may be, are less than L; and where P_{n+1} is obtained from P_n by merely introducing another vertex, P_n is steadily increasing with n. According to Theorem II, regarding monotonic, bounded sequences (Sec. 9), the P_n have a limit—the arc length of C. Now it is easy to show that any inscribed polygon has a perimeter smaller than C and that every circumscribed polygon a perimeter larger than C.

Thus we have clarified what is meant, for example, by the length of a circle, or arc of a circle, or arc of a parabola, etc. How to calculate those lengths is quite a different problem.

3

DIFFERENTIAL
AND INTEGRAL CALCULUS

18. TANGENT PROBLEMS

Tangent problems are easier than area problems. The Greeks knew very well
how to construct a tangent at a point P of a circle (Fig. 63) by drawing the
perpendicular to the radius OP. In the case of the ellipse, the construction of
the tangent ab rested on the theorem that the tangent at P forms equal angles
with the two focal radii drawn from P (Fig. 64). Similarly, in the case of the
hyperbola.

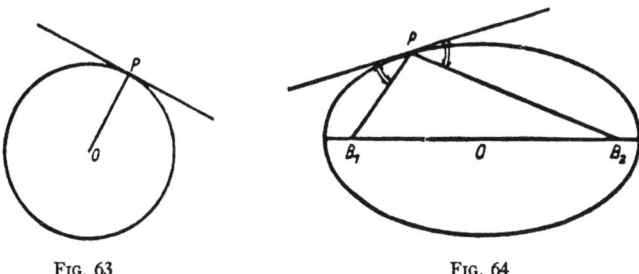

Fig. 63   Fig. 64

(The Greeks, in fact, treated many tangent problems. Archimedes in his
treatise on the spiral[29]—the one which we still call the *Spiral of Archimedes*—
deals with nothing else but the construction of the tangent and the calculation
of the area of a sector of the spiral.)

When in modern times—meaning the first half of the seventeenth century—
Greek mathematics was resumed, numerous new tangent problems were treated.
But, because a curve was now understood as a geometrical representation of a
computational expression, the method of dealing with tangent problems was
changed, and a new element entered into them. For instance, in the case of the
parabola, which is the geometric representation of the function $y = x^2$, we
find the tangent at the point $E(1, 1)$ (Fig. 65). We take a value of x close to 1
and call it x_1; the corresponding value of the function is $y_1 = x_1^2$. Let E_1 be the

point (x_1, y_1); let φ_1 be the angle which the secant EE_1 makes with the horizontal through $E;$ then we have

$$\tan \varphi_1 = \frac{x_1^2 - 1^2}{x_1 - 1} = \frac{f(x_1) - f(1)}{x_1 - 1}.$$

And here is the new method that we wanted to introduce by the example. We regard the tangent as the limiting position of the secant EE_1, which results when x_1 approaches 1 indefinitely.

So much for the principle. With simple mathematical manipulation, we find

$$\frac{x_1^2 - 1}{x_1 - 1} = x_1 + 1.$$

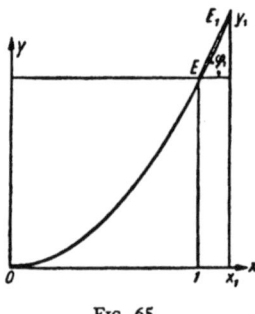
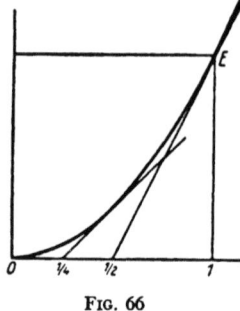

FIG. 65   FIG. 66

If x_1 approaches 1, $x_1 + 1$ approaches 2; this means that if φ is the angle between the tangent and the x-axis

$$\tan \varphi = 2.$$

Thus, in the limit, the opposite side is exactly twice as long as the adjacent side. Hence we need not even draw the parabola to construct the tangent at E. We simply connect with E the midpoint of the segment with ends 0 and 1.

In the same way we find for any other x that, as $x_1 \to x$,

$$\frac{f(x_1) - f(x)}{x_1 - x} = \frac{x_1^2 - x^2}{x_1 - x} = x_1 + x \to 2x.$$

So, for $x = \frac{1}{2}$ we have $\tan \varphi = 1$, which means that the tangent at $(\frac{1}{2}, \frac{1}{4})$ is constructed by connecting $(\frac{1}{2}, \frac{1}{4})$ with $(\frac{1}{4}, 0)$. For drawing the parabola, the knowledge of these two tangents is more useful than the plotting of many of its points (Fig. 66).

In Section 12 we saw how Fermat determined

$$\lim_{x \to 1} \frac{1 - x^{n+1}}{1 - x}.$$

Since

$$\frac{1 - x^{n+1}}{1 - x} = 1 + x + \ldots + x^n$$

for $x \to 1$, this becomes $n + 1$. This same principle gives the general construction of the tangent at any point of the curve $y = x^n$. As $x_1 \to x$,

$$\frac{x_1^n - x^n}{x_1 - x} = x_1^{n-1} + x_1^{n-2}x + \ldots + x_1 x^{n-2} + x^{n-1} \to n x^{n-1}.$$

Next we introduce a new symbol and a new terminology. Quite generally we shall write

$$\lim_{x_1 \to x} \frac{f(x_1) - f(x)}{x_1 - x} = f'(x)$$

and call the value $f'(x)$ the "derivative" of f at x. It is equal to $\tan \varphi$, where φ is the inclination to the x-axis of the tangent (Fig. 67).

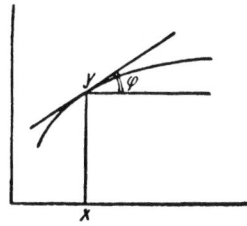

FIG. 67

For these "derivatives" we shall now obtain some very general results, which correspond to the two principles of Cavalieri (see above, Sec. 13).

1. Let $y = f(x) + g(x)$.
Then we have

$$\frac{[f(x_1) + g(x_1)] - [f(x) + g(x)]}{x_1 - x} = \frac{[f(x_1 - f(x)] + [g(x_1) = g(x)]}{x_1 - x},$$

which leads to

$$[f(x) + g(x)]' = f'(x) + g'(x).$$

2. Let $y = \rho f(x)$.
Then we have

$$\frac{[\rho f(x_1) - \rho f(x)]}{x_1 - x} = \rho \frac{f(x_1) - f(x)}{x_1 - x},$$

which leads to

$$[\rho f(x)]' = \rho f'(x).$$

3. For more than two summands these two results give

$$[\rho_1 f_1(x) + \ldots + \rho_n f_n(x)]' = \rho_1 f_1'(x) + \ldots + \rho_n f_n'(x).$$

4. Applying this to the case

$$f_1(x) = 1 , \quad f_2(x) = x , \quad f_3(x) = x^2 , \quad \ldots , \quad f_n(x) = x^n ,$$

$$\rho_1 = c_0 , \quad \rho_2 = c_1, \quad \ldots , \quad \rho_n = c_n ,$$

and considering that

$$f_1'(x) = 0 , \quad f_2'(x) = 1 , \quad f_3'(x) = 2x , \quad \ldots , \quad f_n'(x) = nx^{n-1} ,$$

we obtain

$$[c_0 + c_1 x + c_2 x^2 + \ldots + c_n x^n]' = c_1 + 2c_2 x + \ldots + nc_n x^{n-1} .$$

This means that we can draw tangents at every point of any curve $y = c_0 + c_1 x + c_2 x^2 + \ldots + c_n x^n$.

Fermat, whose achievements in the theory of areas we discussed previously, was a master also in tangent problems. He knew fully how to handle them in situations like those above, as well as for many other curves; therefore, he is often said to have known differential calculus.

19. INVERSE TANGENT PROBLEMS

In contrast with area problems, those concerning tangents presented no difficulties, no matter what curves were taken, until the problem was inverted; that is, the tangent was given and the curve had to be found. In the terminology of functions, which, however, in this form were unknown to the early seventeenth century, this means that, instead of seeking the derivative $f'(x)$ of a given function, $f'(x) = g(x)$ is given, and a function whose derivative is $g(x)$ has to be found. If, for example, $g(x) = c_0 + c_1 x + \ldots + c_n x^n$, we readily find that

$$f(x) = c_0 x + c_1 \frac{x^2}{2} + \ldots + c_n \frac{x^{n+1}}{n+1} + c ,$$

where c is some further constant. If we form $f'(x)$ in accordance with the rules developed in Section 18, c drops out, no matter what its value, and we obtain exactly $g(x)$. Thus we can solve this inverse problem for any polynomial. But, in the case of other functions, difficulties arise immediately.

20. MAXIMA AND MINIMA

Euclid proved that among all rectangles of equal perimeter the square has the largest area. We shall briefly consider his proof, because the difference between his method and the one we are going to use next will repeatedly engage our attention.[30] Figure 68 shows an arbitrary rectangle, together with the square of the same perimeter. Since the perimeter of the rectangle is equal to $2x + 2y + 2u$, and that of the square to $2x + 2y + 2v$, it follows that $u = v$. From the fact that $u = v$, it follows that rectangle A is smaller than rectangle B. For if we regard v as the base of B, and u as the base of A, and with $y = v + x > x$, it appears that A and B have equal bases but that B has a greater height.

That is the Greek proof. Fermat, on the other hand, proceeded quite differently.[31] In Figure 69, let x be one side of the rectangle, a the semiperimeter (wherefore $a - x$ is the length of the other side), and $J = x(a - x)$ its area. For all rectangles of given perimeter the value of a is fixed, x is variable; hence J is a function of x. The value of x can begin at 0 and increase up to a. If x is very small, $a - x$ is almost a; if x is almost equal to a, then $a - x$ is very small. As x varies from 0 to a, the rectangle passes through all possible shapes: from narrow and high, through an "ordinary" one, through the square, to wide and low ones. The question is: For what x does $J(x)$ have the largest value?

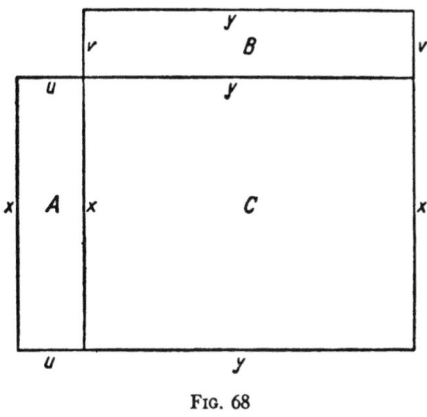

Fig. 68

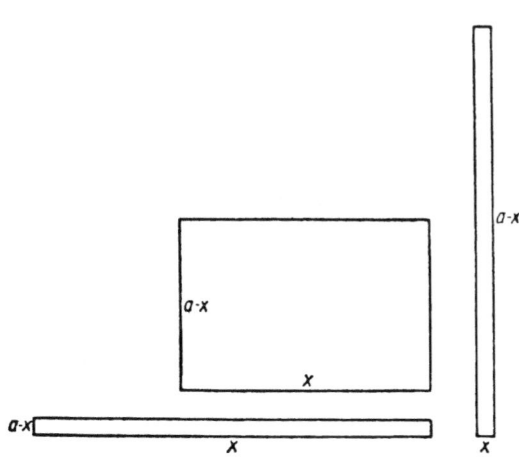

Fig. 69

We represent the function $J(x)$ graphically in the interval $0 \leq x \leq a$ (Fig. 70). At $x = 0$, $J(x)$ begins with 0, and at $x = a$, $J(a) = 0$. We are interested in the highest point, the "maximum" which $J(x)$ attains. Fermat reasons as follows: Let x_0 and x_2 be two values of x such that

$$J(x_0) = J(x_2) .$$

This means

$$x_0(a - x_0) = x_2(a - x_2)$$

or

$$a(x_0 - x_2) = x_0^2 - x_2^2$$

$$a = x_0 + x_2 .$$

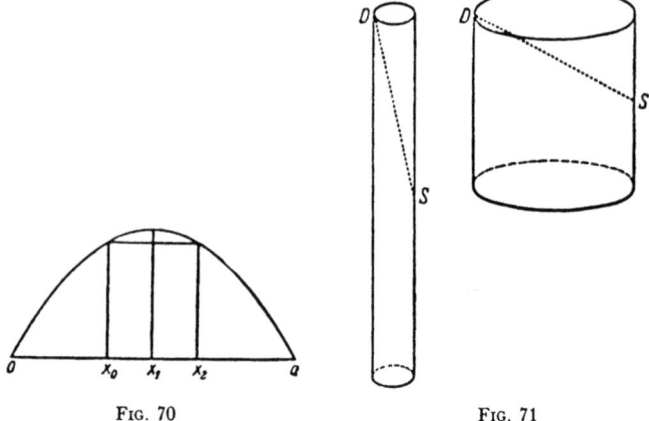

FIG. 70   FIG. 71

As x_0 and x_2 converge, both satisfy ever more closely the condition $x_0 = x_2 = a/2$, the highest point on the curve. But, for $x_1 = a/2$, $a - x_1 = a/2 = x_1$. Hence the rectangle of maximum area is a square.

The same idea could be expressed by saying that at the highest point the tangent must be horizontal, or $J'(x) = 0$. Now, for $J(x) = x(a - x) = ax - x^2$, or $J'(x) = a - 2x$; and, hence, for $a - 2x = 0$, $x = a/2$. The close relation between the two ideas is obvious, although Fermat seems to have felt this only intuitively.

Kepler, too, was strongly aware of this relation, though not in the sense of functions and derivatives. Aside from his astronomical works, an instance of this is to be found in his so-called "Doliometry,"[12] the "barrel calculation." When he, the imperial court astrologer at Linz, married the second time, he bought for the wedding wine from a barrel. To compute the bill, the wine merchant measured the barrel by inserting a foot rule into the taphole S until it reached the lid at D (Fig. 71); then he read off the length $SD = d$ and set the price accordingly. This method outraged Kepler, who saw that a narrow, high barrel might have the same *SD* as a wide one and would indicate the same wine price, though its volume would be ever so much smaller.

Giving further thought to this method of using *d* to determine the volume, Kepler approximated the barrel somewhat roughly by a cylinder, with *v* the radius of the base and *h* the height. Then

$$d^2 = \left(\frac{h}{2}\right)^2 + (2r)^2,$$

or

$$4r^2 = d^2 - \frac{h^2}{4}; \qquad r^2 = \frac{d^2}{4} - \frac{h^2}{16}.$$

Hence, for the volume $V = r^2\pi h$,

$$V = h\pi \left(\frac{d^2}{4} - \frac{h^2}{16}\right) = \pi \frac{d^2 h}{4} - \frac{\pi}{16} h^3.$$

Then he asked: If *d* is fixed, what value of *h* gives the largest volume *V?* *V* is a polynomial in *h;* hence the derivative (though, of course, Kepler did not use derivatives)

$$V'(h) = \pi \frac{d^2}{4} - 3 \frac{\pi}{16} h^2.$$

For *V* to be a maximum, *V'* must equal zero; hence

$$\frac{3\pi}{16} h^2 = \frac{\pi d^2}{4}, \qquad 3h^2 = 4d^2, \qquad h = \frac{2}{\sqrt{3}} d.$$

That defined a barrel of definite proportions. Kepler noticed that in his Rhenish homeland barrels were narrower and higher than in Austria, where their shape was peculiarly close to that having a maximum volume for a fixed *d*—so close, indeed, that Kepler could not believe this to be accidental. So he imagined that centuries ago somebody had calculated barrel shapes, as he himself was doing, and had taught the Austrians to construct their barrels in this particular fashion—a very practical one, indeed. Kepler showed that if a barrel did not satisfy the exact mathematical specification $3h^2 = 4d^2$, but deviated somewhat from it, this would have but little effect on the volume, because near its maximum a function changes only slowly. Thus, while the Austrian method of price determination, if applied to Rhenish barrels, would be a clear fraud, it was quite legitimate for Austrian barrels. The Austrian shape had the advantage of permitting such a quick and simple method. So Kepler relaxed in this instance. Working out finer approximations of various barrel shapes, he consulted Archimedes and discovered that his own method of indivisibles had enabled him to obtain results in a far simpler and more general way than Archimedes, who had been struggling with cumbersome and difficult proofs. What he did not suspect was that Archimedes, too, had found his results by the same method of indivisibles (for the ἔφοδος was lost until 1906!). Kepler devoted to these problems a whole book containing computations of many volumes.

21. VELOCITY

Galileo, Kepler's great contemporary, in a particular instance also arrived at differential calculus and anticipated almost all the reasoning which, more generalized, led to the invention of differential and integral calculus. The idea of a function and its graphical presentation are already clearly present in Galileo (1620); for him they followed quite directly from the old *Mechanics* of Aristotle.

With Euclid-like punctiliousness, he first explained what is meant by uniform motion and velocity. In Figure 72, $v = g$ is a constant, and $s = gt$ the distance traversed in the time interval from $t = 0$ to t, presented as the area of a rectangle of base t and height g. Abruptly departing from Euclidean language and rigor, Galileo claimed this rectangle to be the sum of its verticals, each of them being equal to the velocity v, the distance traversed in unit time; hence their sum is the total distance. But, if the motion in question is not uniform, v is not constant, said Galileo. He assumed it to be proportional to t, $v = gt$. The sum

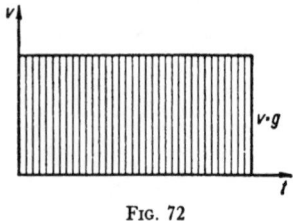

FIG. 72

of all verticals up to the line $v = gt$, that is, the shaded area under that line (Fig. 73), indicates analogously the total distance traversed. In this case, $s = \frac{1}{2}tgt = \frac{1}{2}gt^2$.[33]

(Leonardo da Vinci, trying to solve the problem of the velocity of the falling body, had tentatively placed $s = t^2 + t$. He, too, thus emancipated himself from Aristotle's doctrine that all bodies are falling with uniform velocity, but iron faster than wood. Leonardo, however, was unable to pursue the problem further.)

It has often been claimed that Galileo arrived through experiment at the formula $s = \frac{1}{2}gt^2$ and that he arrived at the theory afterward. It is certainly of great interest to know whether in this first interplay of experiment and theory it was the former which had the leading role. It did not, to be sure. Experimentally to determine the relation between the time and distance of a freely falling body was at that time technically impossible. Only the fact that, contrary to Aristotle, iron and wood do fall equa.ly fast could be ascertained in that way. What made it possible at all for Galileo to approach the problem experimentally was his idea of studying bodies falling on variously inclined planes. Even so, all he could do was to listen to the uniform rhythm with which his rolling balls passed certain markers placed at certain distances. But exactly at what distances? To determine them, he needed an idea, a tentative plan, which could come only from theoretical conjectures. They led him to place the marking lines at distances 1, 4, 9, 16, ... And, in fact, the same rhythm of the rolling ball occurred regardless of the inclination of the plane.

It is unmistakably clear from Galileo's description in the *Discorsi* that first he had the idea and afterward made the experiment. That case became typical of all research in physics. Nobody will ever discover a law by sitting down and merely watching an apparatus nor by idly speculating on nature without ever observing it. It is the correspondence of concepts and experiments the refinement of which makes for all progress in science.

Here we are interested chiefly in the mathematical concepts and processes which Galileo used. What he did was very bold indeed. To assume v, when it is not constant, to be equal to gt was perhaps plausible but nevertheless a cou-

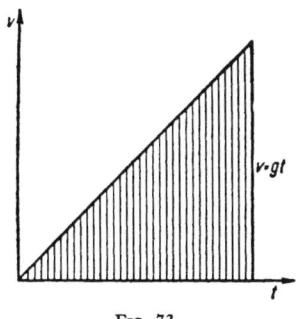

Fig. 73

rageous step. Truly bold, however, was his manner of employing indivisibles in order to arrive at the expression for the distance covered. Only at a later point of our discussion shall we be able fully to appreciate what that means. On the one hand, Galileo envisaged what we now call $_0\int^t gx\,dx$ and would consider to equal $\frac{1}{2}gt^2$. On the other hand, he stated the velocity of this motion $s = \frac{1}{2}gt^2$ to mean $v = gt$. The idea that velocity is the same as a derivative can be read somewhere between the lines. We, of course, are familiar with the fact that gt is the derivative of $\frac{1}{2}gt^2$. The relation of definite integral and derivative seems to have been understood for the first time at this point. So far, however, it was not clearly grasped.

To the "geometrical representation" of a function $y = f(x)$ has thus been added a "mechanical representation." If we interpret $s = f(t)$ as the distance which a point moving along a straight line covers from $t = 0$ to a time t (e.g., a train traveling from the station to its destination), then every function $y = f(x)$ can be interpreted as a motion; and, conversely, any recorded motion, say, that of a train passing milestones, can be described through a "function" $s = f(t)$.

The velocity, however, which for a uniform motion is the constant ratio $(s_1 - s)/(t_1 - t)$, varies from one moment to another when the motion is non-uniform and is describable as a function of t, namely,

$$\lim_{t_1 \to t} \frac{f(t_1) - f(t)}{t_1 - t}.$$

As an area may be defined through a definite integral, so we may say here that the above limit, that is, the derivative, is the definition of the "velocity." Galileo, however, like Aristotle, regarded it as a basic concept without defining it.

22. NAPIER

The needs of astronomy and of navigation demanded ever further refinements of the sine tables and of decimal computation as such, and such simplifications of multiplication, division, and the extraction of roots were prepared in the late sixteenth century by the so-called prosthaphaeretic method based on the increasingly accurate sine tables. The principle was quite simple: From

$$\cos (x + y) = \cos x \cos y - \sin x \sin y$$

$$\cos (x - y) = \cos x \cos y + \sin x \sin y$$

followed

$$\cos (x + y) + \cos (x - y) = 2 \cos x \cos y$$

or

$$\cos x \cos y = \tfrac{1}{2} \cos (x + y) + \tfrac{1}{2} \cos (x - y).$$

To multiply two numbers, A and B, we find in the sine table—which is, of course, also a cosine table—the angles x and y such that $\cos x = A$ and $\cos y = B$, form $x + y$ and $x - y$, take from the table $\cos (x + y)$ and $\cos (x - y)$, and obtain the product AB by addition. This is not bad for the purposes of astronomy and navigation, where sines and cosines are often multiplied, but it was slow; something simpler was needed.

The idea of a table juxtaposing the terms of an arithmetic progression on the left with those of a geometric progression on the right was in the air. Michael Stifel[14] expressed it; Bürgi carried it out, not knowing that Napier had long been working at it. Although Kepler, who was much in need of such tables, had urged Bürgi to get them published, the latter refused to reveal his secret prematurely; "Homo cunctator et secretorum custos," as Kepler put it. As a result, when Napier's tables appeared in print in 1614, Bürgi forfeited any priority for the discovery.[15] As we shall see, his work differed considerably from Napier's, and it is doubtless unjustified to claim for Bürgi priority, or at least a sharing with Napier in the discovery. In fact, he does not deserve such credit.

It is easy to describe Bürgi's approach and very instructive too. We are familiar with using logarithmic, trigonometric, and other tables* in a twofold way: to find the logarithm, etc., of a given number and to find the number having a

---

*Toeplitz assumes a familiarity with these tables that is no longer valid. As this section illustrates, something has been lost in the transition from logarithmic tables to calculators.

given logarithm. What Bürgi constructed (and Napier too) were tables not of logarithms but of "antilogarithms." They contain but the numbers belonging to given logarithms; hence, the procedure of finding the logarithm of a number is inverse. But for practical purposes it is entirely equivalent to our tables.

Bürgi constructs a geometric progression with 10^4 for the first term and $(1 + 1/10^4)$ for the common ratio. To multiply any term by $(1 + 1/10^4)$, however, is the same as adding $1/10^4$ of its value to it. In this way he obtains

$$\begin{array}{r} 10000.0000 \\ 1.00000000 \\ \hline \end{array}$$

$$1.\ \begin{array}{r} 10001.00000000 \\ 1.00010000 \\ \hline \end{array}$$

$$2.\ \begin{array}{r} 10002.00010000 \\ 1.00020001 \\ \hline \end{array}$$

$$3.\ \begin{array}{r} 10003.00030001 \\ 1.00030003 \\ \hline \end{array}$$

$$4.\ \ 10004.00060004$$
$$\cdots$$

It is true that in the computation shown here, something questionable happens: In the third step we ought, in fact, to add 1.000300030001, that is, a little more! But Bürgi—as Napier—omitted those higher places. The true value lies therefore never below but always above the abbreviated one; in fact, in each step $1/10^4$ of the last place which is carried (the fourth after the decimal point) is dropped. After 10,000 steps this error would amount to one unit in the last place.

How far does Bürgi continue his progression? To 23,027 terms! That is a big computing job, but each step is very simple. A single mistake, however, would ruin all that follows. The 23,027th step takes him close to 100,000; the next step would take him past 100,000. There he stops.

To illustrate the use of these tables, we compute $\sqrt{36,000}$. For the step which gives 36,000, the table shows

$$\begin{array}{rl} 12809\} & 35996.4763 \\ 12810\} & 36000.0759. \end{array}$$

"Interpolation" then gives 12809.98 as the exact step (or place) to which 36.000.0000 belongs. Dividing this by 3, we obtain 4,299.99. The number belonging to the 4,299.99th step is $\sqrt[3]{36,000}$.

Such a table certainly is equivalent to our modern four-place tables of logarithms. For the scientific purposes of those days, however, four-place tables were not at all accurate enough; rather seven-place tables were needed. It would be difficult, however, to "transform" Bürgi's table into a seven-place one. For if we began with 10^7, and, by using the ratio $(1 + 1/10^7)$, obtained

$10^7(1 + 1/10^7)^n$ in the nth step, the numbers of the sequence, while analogous to the first, would be totally different numbers. Let us demonstrate this by constructing a very small table, using $10(1 + 1/10)^n$ as the nth term, which consists of only twenty-four numbers and shows, of course, altogether different numbers:

$$10.00$$
$$1.0000$$

1. 11.0000
      1.1000

2. 12.1000
      1.2100

3. 13.3100
      1.3310

4. 14.6410
      $. . .$
      $. . .$

24. 98.4948
      9.8494

25. 108.3442

But what if Bürgi had used $10^7(1 + 1/10^7)^n$ as the nth step?

$$10\ 000\ 000.000\ 000$$
$$1.000\ 000$$

1. $10\ 000\ 001.000\ 000$
      $1.000\ 000$

2. $10\ 000\ 002.000\ 000$
      $. . .$

This would take about 23,000,000 steps, which is practically impossible to carry out.

Napier, in fact, started out doing just this; he carried the calculation out for 100 steps. But then he conceived an idea how to substitute reasoning for mechanical computation. It is this idea which interests us here, not logarithms as such. It is this concept that is Napier's great achievement, while the construction of the sequence as such, which he had in common with Bürgi, was, so to speak, in the air in those days. Napier did not publish his idea in 1614 together with his tables and the directions for their use, but his son did in 1619 after Napier's death in 1617.

First of all, Napier clearly realized the relationship between all tables, regardless whether they use as this starting point $(1 + 1/10)^n$, or $(1 + 1/10^4)^n$, or any other such ratio. The essential thing to him was that the numbers in the table form a geometric progression paired with the arithmetic progression of the numbered steps: If a is a number belonging to step α, and b a number belonging to step β, then ab is the number belonging to step $\alpha + \beta$. In our 24-number table above the number 100 belongs to step 24.153; in Bürgi's table the number 100,000 belongs to step 23.027. On the other hand, in our table the number 36 belongs to step 13.428, while in Bürgi's table 36,000 belongs to step 12.810. Now

$$\frac{23.027}{12.810} = 1.798, \text{ and } \frac{24.153}{13.428} = 1.799 ,$$

that is, within the limits of computational accuracy, the two ratios are equal. This means that the step-numbers in the one table can be found from those of the other through multiplication by the fixed factor 23.027/24.153.

This can also be shown in a more general way. Let

$$1 + \frac{1}{10} = A , \quad 1 + \frac{1}{10^4} = B. \quad \text{Also } A^\alpha = 10, \quad B^\beta = 10, \quad A^\gamma = a, \quad B^\delta = a .$$

Then we have

$$A^\alpha = B^\beta , \quad A^\gamma = B^\delta , \quad \text{or} \quad (A^\alpha)^{\gamma/\alpha} = (B^\beta)^{\delta/\beta}$$

and, because of $A^\alpha = B^\beta$,

$$\frac{\gamma}{\alpha} = \frac{\delta}{\beta}, \quad \alpha : \beta = \gamma : \delta .$$

Accordingly, Bürgi's table might be transformed by the following steps. First, compute the table $10^2(1 + 1/10^2)^n$, which contains about 230 tabulations. Second, transform these by the use of the proper factor into those of the table $10^4(1 + 1/10^4)^n$; this gives every hundredth entry of the Bürgi table. Third, compute the first hundred steps, as Bürgi had done, and fill up the intervals between the 100th, 200th, 300th, . . . , number.

Napier used this principle for constructing his tables and thereby achieved the otherwise unmanageable amount of computational work. But he failed to give directives how to find the numbers α and β. For $10^2(1 + 1/10^2)^n$ this number results from the process of computation. But how find Bürgi's value 23.027 without going through with the computation of his whole table? Napier thought it unwise to concentrate efforts on any one of these tables, which might have to be abandoned if higher accuracy were required. Thus he searched among all the possible tables for an *absolute* one.

The fundamental property of the logarithm is that it is a function $f(x)$ varying in arithmetic progression when x varies in geometric progression: $f(x), f(\rho x),$ $f(\rho^2 x), f(\rho^3 x), \ldots ,$ must form an arithmetic progression. Function $f(x)$ is to be such a function that $f(xy) = f(x) + f(y)$. Two numbers are multiplied by adding their logarithms and then by taking the antilogarithm. This means that, for $x = y = 1$,

$$f(1) = f(1) + f(1) = 2f(1) ;$$

hence (1)

$$f(1) = 0 \ ;$$

and, further,

$$\frac{f(x_1) - f(x)}{x_1 - x} = \frac{f[(x_1/x)x] - f(x)}{(x_1/x)x - x} = \frac{f(x_1/x) + f(x) - f(x)}{x[(x_1/x) - 1]}$$

$$\frac{f(x_1) - f(x)}{x_1 - x} = \frac{1}{x} \frac{f(x_1/x)}{(x_1/x) - 1} \ .$$

Without using these formulas explicitly, Napier inferred two other consequences which constitute the mathematical core of his theory:

(2) If x_1 approaches x, the left side approaches the derivative of $f(x)$. (Napier thought of x and $f(x)$ as two points, one of which was moving along a straight line with uniform velocity, the other along another straight line with non-uniform velocity. That is, he expressed by "velocity" what we mean by the derivative.) But, while the left side approaches $f'(x)$, the factor $1/x$ on the right side remains unchanged. The factor

$$\frac{f(x_1/x)}{(x_1/x) - 1} \ ,$$

on the other hand, approaches a limit which is independent of $x;$ for if $y = \rho x$ is some other value and if $y_1 = \rho x_1$ is a neighboring value, then

$$\frac{f(y_1/y)}{y_1/y - 1} = \frac{f(\rho x_1/\rho x)}{(\rho x_1/\rho x) - 1} = \frac{f(x_1/x)}{(x_1/x) - 1} \ .$$

Hence the two limits are the same. Let

$$\lim_{x_1 \to x} \frac{f(x_1/x)}{(x_1/x) - 1} = c \ .$$

Then

$$f'(x) = \lim_{x_1 \to x} \frac{f(x_1) - f(x)}{x_1 - x} = c \ \frac{1}{x} \ .$$

All possible tables of logarithms, therefore—for this has been proved—differ from one another merely by a constant factor. (This has been proved above under a special assumption only.) Now Napier says that among all possible logarithmic functions he wants to take the one for which $c = 1$; let it be designated by log x. Hence

$$(\log)' = \frac{1}{x}$$

(3) The comparison of the three functions $(x - 1)/x$, log x, and $x - 1$ shows that all of them are zero for $x = 1$ and that they all increase from there on. Their derivatives are, respectively, $1/x^2$, $1/x$, and 1 (Fig. 74), and, since for

$x > 1$ the inequalities hold $1/x^2 < 1/x < 1$, the first curve rises slower than log x, and the third rises more rapidly; hence

$$\frac{x-1}{x} < \log x < x - 1$$

or

$$\frac{1}{x} < \frac{\log x}{x-1} < 1$$

or, placing $x = a/b$, $(a > b)$

$$\frac{b}{a} < \frac{\log (a/b)}{(a/b) - 1} < 1$$

or

$$\frac{1}{a} < \frac{\log a - \log b}{a - b} < \frac{1}{b}, \qquad \text{if } a > b.$$

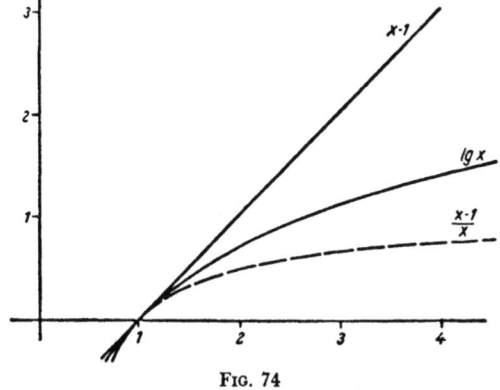

FIG. 74

Napier used this basic formula for a double purpose. First, he used it to compute the factor which transforms, for example, the table $10^2[1 + (1/10^2)]^n$ into his, Napier's, logarithms: for $a = 1 + (1/10^2)$, $b = 1$, the first two entries in the table, $1/101 < \log 1.01/1 < 1/100$. Second, he used the formula for purposes of ever necessary interpolation; it saved him the labor of recurring divisions.

Actually, in Napier's presentation all this is complicated by a number of circumstances ultimately irrelevant to his central idea. But, because of them, this central idea has never been clearly presented disentangled from all these minor contingencies. Napier never tabulated log x but log sin x; moreover, he did not add in the tables but subtracted; and, finally, there is a computational error in his Table II owing to which his tables are wrong beyond the seventh decimal. Other accounts of his work rather stress these shortcomings, but I am inclined to disregard them in order to show the essence of his reasoning.

We wonder how Napier hit upon all these ideas so alien to the then generally accepted framework of Euclid's geometrical methods and so modern in some elements. Unfortunately, many of his manuscripts were destroyed by fire. His great-grandson Mark Napier wrote a big book about him[36] with the purpose of showing that Napier, a Scotch baron who spent most of his life in a lonely castle, had conceived all his ideas from his own thinking. Actually, he did a real disservice to his ancestor, for the greatness of Napier's own ideas would appear only if we could compare them with those of his predecessors. It is known that as a young man he traveled in France for a long time, and it is likely that in 1594 he spent some time in Italy, especially in Padua, where Galileo was then teaching. But that is all we know. We do not know where he made contact with the theories on motion and functions, which can be traced from Aristotle to Galileo.

Upon close study there appears another noteworthy feature in Napier's work. He did not simply disregard the higher decimals but was reasoning whether the dropping of higher decimals raised or diminished the true value; that is, he strove to be strictly exact in his statements. But then there is a place in his computations where he abandoned this strictness; obviously, his principal formula did not permit him to obtain accurate enough values for his interpolations. In speculating as to why Napier never published his *constructio*, authors seem to have overlooked this most natural motive that he was dissatisfied with these less exact parts of his work and was hoping to perfect it.

When Briggs, professor of mathematics in London, became acquainted with Napier's tables in 1614, he broke off his regular course and instead lectured on logarithms. As soon as the semester was over, he went to Scotland and spent a long time with Napier. He persuaded Napier—which was not difficult—that it would be better to take logarithms to a base of 10, that is, those which result from the inversion of the function $y = 10^x$.

So far we have not spoken of bases. The question is: For what q is $y = q^x$ the inverse of $x = \log y$? If in Napier's formula we place $a = 1 + 1/n$, $b = 1$, we obtain

$$\frac{1}{1 + 1/n} < \frac{\log(1 + 1/n) - \log 1}{(1 + 1/n) - 1} < \frac{1}{1},$$

or

$$\frac{n}{n+1} < \frac{\log(1 + 1/n)}{1/n} < 1,$$

or

$$\frac{1}{n+1} < \log\left(1 + \frac{1}{n}\right) < \frac{1}{n},$$

or

$$\frac{n}{n+1} < n \log\left(1 + \frac{1}{n}\right) < 1.$$

or

$$\frac{n}{n+1} < \log\left(1 + \frac{1}{n}\right)^n < 1.$$

For $n \to \infty$, because of

$$\lim_{n \to \infty} \left(1 + \frac{1}{n}\right)^n = e \text{ (Sec. 7)};$$

therefore,

$$1 \leq \log e \leq 1 ;$$

that is,

$$\log e = 1 ; \quad \log (e^x) = x \log e = x .$$

Hence the base q of the inverse of the function $y = \log x$ is the number e which we encountered before. Now, why should it be more practical to use logarithms to the base 10—which we shall designate by log x? A simple numerical example makes this clear. Let us compute 169^2 with Napier's table. The table gives log x for $x = 100, 101, 102, \ldots, 1,000$. For example,

$$\log 169 = 5.12990 ;$$

hence

$$\log 169^2 = 2 \log 169 = 10.25980 .$$

But the table does not go that far; it ends with log $1,000 = 6.90776$. Hence we proceed as follows:

$$
\begin{array}{r}
\log 169^2 = 10.25980 \\
(-)\log 100 = 4.60517 \\
\hline
\log \dfrac{169^2}{100} = 5.65463 .
\end{array}
$$

The integer whose logarithm is closest to 5.65463 is 286. Hence

$$\frac{169^2}{100} \sim 286, \quad 169^2 \sim 28,600 .$$

Using the base 10, we have

$$\log 169 = 2.22789$$

$$\log 169^2 = 4.45578 .$$

The table does not contain this logarithm but only 0.45578. Now, instead of a cumbersome subtraction, we account for the integer difference in the logarithms by merely shifting the decimal point. Hence for purposes of numerical computation the base 10 is more practical.

Briggs,[37] it is true, together with Napier, recomputed the whole table for log x (base 10) quite independent of Napier's earlier computations (if only because of the numerical error in Napier's tables). The new computation interestingly shows a new application of Napier's fundamental formula. Above, we have derived from it the inequality

$$\frac{1}{n+1} < \log \left(1 + \frac{1}{n}\right) < \frac{1}{n} .$$

Applied for example, to $n = 2,400$:

$$\frac{1}{2,401} < \log\left(1 + \frac{1}{2,400}\right) < \frac{1}{2,400}.$$

The error for log 2,401/2,400 is less than $1/2,400 - 1/2,401 \sim 1/5,760,000$, which is equivalent to a seven-place accuracy. What is gained thereby?

$$2,401 = 7^4, \qquad 2,400 = 2^5 \cdot 3 \cdot 5^2 ;$$

hence

$$\log \frac{2,401}{2,400} = \log 2,401 - \log 2,400 = 4 \log 7 - 5 \log 2 - \log 3 - 2 \log 5 .$$

Once log 2, log 3, and log 5 are computed, log 7 can easily be obtained up to seven decimal places. This trick can be used to compute successively log 2, log 3, log 5, log 7, log 11, . . . —in brief, the logarithms of all prime numbers and, since log 6 = log 2 + log 3, etc., also the logarithms of all natural numbers. Next, because of log (p/q) = log p — log q we can compute also those of all fractions, in particular those whose denominator is 10^n, and thus the tables for log x.

When Kepler received Napier's tables, published in 1614, he needed them most desperately, for he was then engaged in the enormous computations which led him to the discovery of his third and last law of planetary motions. Full of enthusiasm over this new device he wrote his old teacher Maestlin in Tübingen about it. The latter, used to the "prostapheretic" method, in reply (March, 1620) made the following significant comment:

I can see that the logarithmic computation is providing correct results, but I am not going to use it. For so far I have been unable to figure out how the tables are constructed, which makes me suspect that the inventor, on purpose, used some obstruse number as the base to make it difficult, if not impossible to check it. I regard it as unworthy of a mathematician to see with other people's eyes and to accept as true or as proven that for which he himself has no proof. For one may always doubt whether these computations, even though correct ten or a hundred times, may not one day provide wrong results.[38]

The important thing is Kepler's reaction to this blend of conservatism and wisdom. Since he did need the tables, and since Napier's *constructio* was not available, he himself began to compute tables. And, great mathematician that he was, he at once discovered Napier's fundamental formula and a dozen or more better ones as well, of which we mention, as an example (cf. Exercise 7)

$$\frac{\log a - \log b}{a - b} < \frac{1}{\sqrt{ab}},$$

which leads at once to

$$\log\left(1 + \frac{1}{n}\right) < \frac{1}{n}\frac{1}{\sqrt{1 + (1/n)}} = \frac{1}{\sqrt{n(n+1)}} < \frac{1}{n}.$$

23. THE FUNDAMENTAL THEOREM

The discoveries which we discussed in the several sections of this chapter contain the germs of the ideas which led to the invention of the differential and integral calculus. And this development throws a bright light on that point which the customary accounts usually slight if not deliberately hide, even though it is the only new and surprising idea in the whole story. Let us recapitulate.

Fermat knew some parts of the differential calculus, as well as solutions of certain inverse problems. He was also in possession of $^b\!\int_a x^n dx$. Gregorius a Santo Vincentio discovered that $f(t) = {}^t\!\int_1 (dx/x)$ has the property $f(uv) = f(u) + f(v)$. We now notice—which in Gregorius' work was also not noticed until after its publication—that so far nothing had been said about which logarithm this is. We shall now elucidate this. In Figure 75, $f(t)$ is the shaded

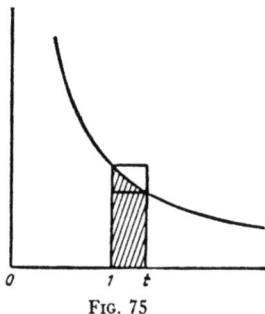

Fig. 75

area. According to the theorems on definite integrals—and also immediately obvious from the figure—we have

$$\frac{1}{t}(t-1) < \int_1^t \frac{dx}{x} < 1\,(t-1).$$

The derivative at $t = 1$, however, is

$$f'(1) = \lim_{t \to 1} \frac{f(t) - f(1)}{t-1}.$$

But, since

$$f(1) = \int_1^1 \frac{dx}{x} = 0,$$

this leads to

$$\frac{f(t) - f(1)}{t-1} = \frac{f(t)}{t-1};$$

and, since

$$\frac{1}{t} < \frac{f(t)}{t-1} < 1,$$

$$\lim_{t \to 1} \frac{f(t)}{t-1} = f'(1) = 1.$$

This means that $f(t) = \log t$, the natural or Napier's logarithm—a fact which additionally justifies the use of this logarithm. Hence we now have

$$f'(t) = \frac{1}{t}.$$

Now let us summarize the facts:

1. $\qquad F(t) = \int_a^t x^n dx = \frac{t^{n+1}}{n+1} - \frac{a^{n+1}}{n+1},\qquad F'(t) = t^n,$

2. $\qquad F(t) = \int_1^t \frac{dx}{x} = \log t,\qquad F'(t) = \frac{1}{t}.$

Let us finally recall Galileo's bold reasoning:

$$F(t) = s(t) = \int_0^t gx\,dx = g\,\frac{t^2}{2},$$

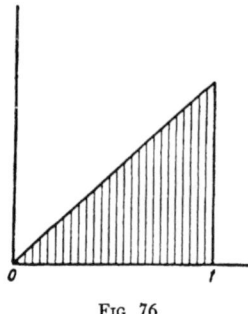

Fɪɢ. 76 Fɪɢ. 77

where the velocity v for $s = g(t^2/2)$ is $v = gt$. That is a special case of 1, for $n = 1$. In the derivation of this relation, however, a connection had been seen between these two facts: $s(t)$ was regarded as the sum of the verticals each of which represented the velocity at a moment t of the motion (Fig. 76). By 1650 all these facts were on hand. But only a man of the next generation, Isaac Barrow, recognized in 1667 the basic relation which appeared here—the *fundamental theorem:*

If $F(t) = {}^t\!\int_a f(x)dx$, $a \leq t \leq b$, *and if* $f(x)$ *is "continuous" and monotonic for* $a \leq x \leq b$, *then* $F'(t) = f(t)$.

In other words, the derivative of $F(t)$ is the function under the integral sign, the "integrand"; or, yet in other words, the definite integral regarded as a function of the upper limit solves the inverse problem of finding a function whose derivative is the given function $f(x)$.

Barrow's proof is both simple and rigorous (Fig. 77). We consider

$$F'(t) = \lim_{t_1 \to t} \frac{F(t_1) - F(t)}{t_1 - t};$$

however,

$$F(t_1) - F(t) = \int_a^{t_1} f(x)\,dx - \int_a^t f(x)\,dx = \int_t^{t_1} f(x)\,dx$$

is the area of the shaded figure. Since $f(x)$ is supposed to be monotonic, we have

$$f(t) < f(x) < f(t_1) \qquad \text{for} \qquad t < x < t_1,$$

and

$$(t_1 - t)f(t) < \int_t^{t_1} f(x)\,dx < (t_1 - t)f(t_1);$$

hence

$$f(t) < \frac{F(t_1) - F(t)}{t_1 - t} < f(t_1).$$

For $t_1 \to t$, $f(t_1)$ passes through all functional values from t_1 to t. Should $f(x)$ take a jump at $x = t$ (Fig. 78), it would still be monotonic, but it would no

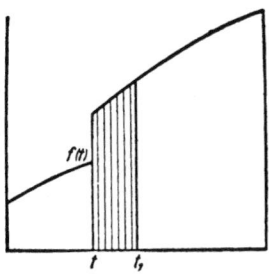

FIG. 78

longer be true that $\lim_{t_1 \to t} f(t_1) = f(t)$. Now, we speak of a function $f(x)$ as being "continuous" at $x = t$ if $\lim_{t_1 \to t} f(x) = f(t)$ regardless of whether x approaches t from the left or from the right. Since $f(x)$ had been supposed to be "continuous," it follows that

$$\lim_{t_1 \to t} \frac{F(t_1) - F(t)}{t_1 - t} = f(t).$$

Now we also recognize that this relation had been intuitively anticipated by Galileo. For his reasoning which he applied to $f(x) = c$ and $f(x) = gx$ is in fact generally applicable; to him $F(t)$ is not the sum of small rectangular strips but the sum of the individual verticals. These to him are, on the one hand, indivisible elements of the area, infinitely narrow strips whose area is simply given by their height; on the other hand, their height is the velocity. This reasoning contains the essence of Barrow's discovery and proof. If Archimedes had been the discoverer of all this, he would have found it as Galileo did and then proved it as Barrow did, deriving it from explicitly and sharply formulated assumptions.

Thus the connection had been established between the theory of the definite integral, on the one hand, and the differential calculus, on the other. In his book[39] Barrow developed a great many theorems of the differential calculus which Fermat had known twenty years earlier, without publishing them, and the ground for which had been prepared by many other mathematicians (Galileo, Torricelli, Cavalieri, Roberval, Pascal).

Why, then, is it that Barrow is not credited with being the discoverer of the infinitesimal calculus but Newton and Leibniz are, between whom that unhappy priority dispute then started to rage? This question we had better postpone until we are better acquainted with the differential calculus of Newton and Leibniz, which is still our present calculus. This much, however, we wish to say right here: In a very large measure Barrow is indeed the real discoverer—insofar as an individual can ever be given credit within a course of development such as we have tried to trace here. Yet there was something lacking in his work. To understand what it was that was lacking—not any factual knowledge, for sure—we need a careful critical analysis which we shall be able to undertake only when the whole subject has been fully developed.

First, however, we must recast the fundamental theorem into another, different form which brings out a new relation. In the formulation we have given so far, the differential calculus was the gainer; if we have found the definite integral $\int_a^t f(x)dx = F(t)$, we have thereby found a function whose derivative is $f(t)$, the so-called indefinite integral, $\int f(x)dx$, in Leibniz' notation. What, then, if we possess an indefinite integral $\varphi(x)$ of a function $f(x)$, that is, a function $\varphi(x)$ such that $\varphi'(x) = f(x)$? Could there be more than one function whose derivative is $f(x)$?

Let us answer these questions, the last one first. If $\varphi(x)$ is the so-called indefinite integral, $\varphi'(x) = f(x)$, then $\varphi(x) + c$ is likewise an indefinite integral, since $[\varphi(x) + c]' = \varphi'(x) + 0 = \varphi'(x) = f(x)$ for any c. This means that, if there is one function which has $f(x)$ for its derivative, then there are infinitely many functions. Yet, if $f(x)$ is continuous and monotonic, then, according to the fundamental theorem, there is indeed one such function, $F(t)$. But does $F(t) + c$ represent the totality of functions which are solutions of our problem? Yes, it does! For assume that besides $\varphi(x)$ there were some other solution $\psi(x)$, such that $\varphi'(x) = f(x)$ and $\psi'(x) = f(x)$ for all $a \leq x \leq b$. Let $g(x) = \psi(x) - \varphi(x)$. Then $g'(x) = \psi'(x) - \varphi'(x) = f(x) - f(x) = 0$. That is, $g(x)$ is a function whose derivative is everywhere zero. But now there is the other fundamental theorem which asserts:

If the derivative of a function is zero for all x in the interval a $\leq$ x $\leq$ b, *then the function is a constant.*

Geometrically, this is immediately obvious. If the curve is horizontally directed at every point, it does represent a constant (Fig. 79). And it is even more obvious—if we may use such an expression—if the function is interpreted in terms of motion. For the derivative to be zero means that the velocity is zero; the body is at rest. But if a moving point is at rest at every moment, it stands still, and its place s is constant.

Hence, from this theorem, whose rigorous mathematical proof we postpone until later, it follows indeed that

$$g(x) = c \; ; \quad \psi(x) - \varphi(x) = c \; ; \quad \psi(x) = \varphi(x) + c \, .$$

All solutions of the problem $\varphi'(x) = f(x)$ *result in a single function* $F(x)$ *in the form* $\varphi(x) = F(x) + c$.

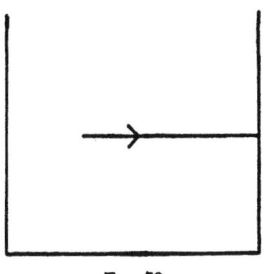

FIG. 79

And now we can solve the problem of finding $^b\!\int_a f(x)dx$ if we have $\varphi(x)$. We simply form $\varphi(b) - \varphi(a)$. For, if $\varphi(x) = F(x) + c$, where $F(t) = {}^t\!\int_a f(x)dx$, we have

$$\varphi(b) = F(b) + c = \int_a^b f(x)\,dx + c$$

$$\varphi(a) = F(a) + c = \int_a^a f(x)\,dx + c = c$$

$$\overline{\varphi(b) - \varphi(a) = F(b) - F(a) = \int_a^b f(x)\,dx\,.}$$

If for a given function $f(x)$ *an indefinite integral* $\varphi(x)$ *is known, then* $^b\!\int_a f(x)dx = \varphi(b) - \varphi(a)$.

For polynomial functions this rule had been obtained already in chapter ii, Section 13.

24. THE PRODUCT RULE

If $u(x)$ and $v(x)$ are two functions of x, we obtain $\underline{(uv)' = u'v + uv'}$. In this case, too, it would be interesting to tell how it was discovered; but here lies the line between history for its own sake and history for the sake of illuminating the development of mathematical thought. The history of the discovery of the fundamental theorem served to illuminate aspects which usually do not stand out clearly at all, and this could hardly have been achieved by any other method. About the product rule, however, there is nothing to be illuminated. Hence the history of its discovery does not concern us.

Other things need to be clarified here. Before proving this first rule of differentiation, and thereby opening the whole procession of all these rules, we should clarify the *definition of the derivative.* Just as for the definition of the limit and of the definite integral, it is a "prescription for a definition" rather.

If $\lim\limits_{x_1 \to x}$ [f(x₁) − f(x)]/(x₁ − x) = *a exists, regardless of whether* x_1 *approaches* x *from the left or from the right, we say that* f(x) *"is differentiable at* x" *or that "it has a derivative* a = f′(x) *at* x."

We shall appreciate this definition by drawing a figure in which it fails. We draw a curve which has a "kink" at *x* (Fig. 80). It has a tangent at *P* from the

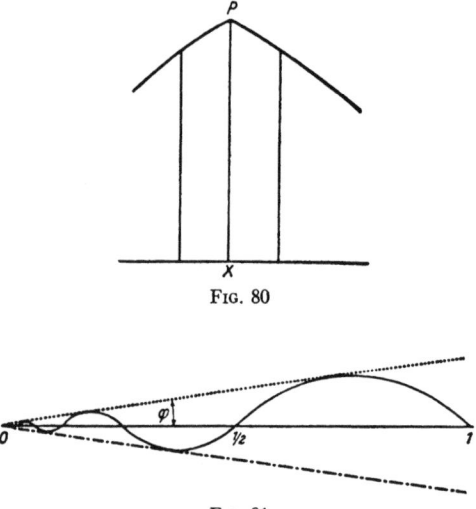

FIG. 80

FIG. 81

left and from the right; but the two are not identical, as they usually are. But there is also another way in which the definition may fail; it can happen that the limit, taken from the right only, simply does not exist. To demonstrate this, we draw (Fig. 81) over the segment from $\frac{1}{2}$ to 1 an arc of a curve, say, a quarter of a circle, which is inclined by 45° at $\frac{1}{2}$ and at 1. Next between $\frac{1}{2}$ and $\frac{1}{4}$ we draw a quarter-circle half as large in radius, lying below the horizontal; next between $\frac{1}{4}$ and $\frac{1}{8}$, again above the horizontal, and so on, alternatingly above and below. The function $f(x)$, which for $0 < x \le 1$ is thus everywhere defined, has a tangent everywhere in the interval $0 < x \le 1$. As to $x = 0$, we have so far not even defined what $f(0)$ is. But we see that

$$\lim_{x \to 0} f(x) = 0 .$$

Hence, if we define $f(0) = 0$, $f(x)$ is continuous at $x = 0$. Is the function also differentiable at $x = 0$? It is not! For from the theory of similar figures it follows that all upper circular arcs are tangent to the dotted line and all lower ones to the dash-dotted line. Thus, as $x \to 0$, the quotient

$$\frac{f(x_1) - f(0)}{x_1 - 0}$$

will alternatingly become tan φ, where φ is the angle of inclination of the dotted line, and next $-\tan \varphi$, forever alternating, with φ being constant while x_1 varies. The difference quotient thus oscillates between tan φ and $-\tan \varphi$, without tending toward a limit. The same would be true on the left side of the figure if we define $f(x)$ for $-1 \leq x < 0$ in terms of mirror symmetry.

Hence, a function, even though continuous, may fail to have a derivative at a point in several ways. In Figures 80 and 81 $f(x)$ was continuous both at the point in question and elsewhere. This shows that a function, although continuous at x, need *not* be differentiable at x. The converse, however, is true, and this is what we need here:

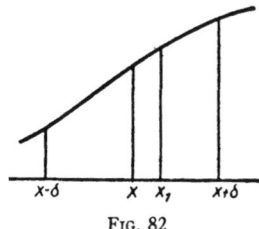

FIG. 82

If f(x) *is differentiable at a point* x, *then it is continuous at* x.

For if

$$\lim_{x_1 \to x} \frac{f(x_1) - f(x)}{x_1 - x} = f'(x)$$

exists, then (according to the theorem that a convergent sequence is bounded),

$$\left| \frac{f(x_1) - f(x)}{x_1 - x} \right| \leq M$$

for a certain interval around x, from $x - \delta$ to $x + \delta$ (Fig. 82). Hence

$$|f(x_1) - f(x)| \leq M |x_1 - x| .$$

Now, if $x_1 \to x$, the right side approaches zero; hence also the left side; that is,

$$\lim_{x_1 \to x} f(x_1) = f(x),$$

that is, $f(x)$ is continuous at x.

Now we are ready for a neat proof of the product rule and also for the complete formulation of that rule.

1. *If* u(x) *and* v(x) *are both differentiable at* x, *their product is likewise differentiable at* x *and is given by* (uv)$'$ = u$'$v + uv$'$.

Proof. Let $w(x) = u(x)v(x)$. Then

$$\frac{w(x_1) - w(x)}{x_1 - x} = \frac{u(x_1)v(x_1) - u(x)v(x)}{x_1 - x}$$

$$= \frac{u(x_1)v(x_1) - u(x)v(x_1)}{x_1 - x} + \frac{u(x)v(x_1) - u(x)v(x)}{x_1 - x}$$

$$= \frac{u(x_1) - u(x)}{x_1 - x} v(x_1) + \frac{v(x_1) - v(x)}{x_1 - x} u(x).$$

If now x_1 approaches x from either left or right, the only question concerns $v(x_1)$. However, since $v(x)$, according to hypothesis, is differentiable at x, it is— as just proved—also continuous at x; hence

$$\lim_{x_1 \to x} v(x_1) = v(x)$$

and hence

$$\lim_{x_1 \to x.} \frac{w(x_1) - w(x)}{x_1 - x}$$

$$= \lim_{x_1 \to x} \frac{u(x_1) - u(x)}{x_1 - x} \cdot \lim_{x_1 \to x} v(x_1) + \lim_{x_1 \to x} \frac{v(x_1) - v(x)}{x_1 - x} \cdot u(x)$$

$$= u'(x) \qquad \cdot v(x) \qquad + v'(x) \qquad \cdot u(x).$$

Hence the limit exists and has the asserted value.

2. *If* $u(x)$ *is differentiable and* $u(x) \neq 0$, *then* $1/u(x)$ *is also differentiable at* x_1, *and* $(1/u)' = -u'/u^2$.

In that case, namely, we have

$$\frac{[1/u(x_1)] - [1/u(x)]}{x_1 - x} = \frac{[u(x) - u(x_1)]/[u(x)u(x_1)]}{x_1 - x}$$

$$= -\frac{u(x_1) - u(x)}{x_1 - x} \cdot \frac{1}{u(x)} \cdot \frac{1}{u(x_1)}.$$

As x_1 approaches x, this approaches, as before, $-u'/u^2$. We observe that it is quite safe to divide by $u(x_1)$, since the postulated continuity of $u(x)$ assures us that, for x_1 sufficiently close to x, $u(x_1) \neq 0$.

3. *If* $u(x)$ *and* $v(x)$ *are both differentiable at* x, *and if* $v(x) \neq 0$, *then* $u(x)/v(x)$ *is likewise differentiable at* x, *and* $(u/v)' = (u'v - uv')/v^2$.

The proof follows from Rules 1 and 2:

$$\left(\frac{u}{v}\right)' = \left(u \frac{1}{v}\right)' = u' \frac{1}{v} + u \left(\frac{1}{v}\right)' = \frac{u'}{v} - u \frac{v'}{v^2} = \frac{u'v - uv'}{v^2}.$$

We are now at once able to differentiate all rational functions

$$f(x) = \frac{a_0 + a_1 x + \ldots + a_n x^n}{b_0 + b_1 x + \ldots + h\ x^q}.$$

The derivative is another rational function. In particular, we obtain

$$\left(\frac{1}{x^n}\right)' = -\frac{nx^{n-1}}{x^{2n}} = -nx^{-n-1} \quad \text{or} \quad (x^{-n})' = (-n)x^{(-n)-1},$$

that is, the rule for $(x^n)' = nx^{n-1}$ holds for negative n, too.

25. INTEGRATION BY PARTS

If $\varphi'(x) = f(x)$, that is, if $\varphi(x)$ is an "indefinite integral of $f(x)$," then write, following Leibniz, $\varphi(x) = \int f(x)dx$. This notation differs from the usage of mathematicians inasmuch as, according to it, $\varphi(x) + c$ is likewise equal to $\int f(x)dx$. From this, we should not conclude that $c = 0$; but we should simply keep in mind that "$= \int f(x)dx$" always means "$= F(x) + c$," as long as the integral sign has no indicated limits. It would be easy enough for us to avoid this imprecise notation, but then the student would not learn to read it with proper caution in books, all of which use it.

The rule $(uv)' = u'v + uv'$ can then also be written

$$uv = \int u'vdx + \int uv'dx,$$

or

$$\int u'vdx = uv - \int uv'dx.$$

In particular we obtain from this in exact notation

$$\int_a^b u'v\,dx = [uv]_a^b - \int_a^b uv'\,dx,$$

where

$$[uv]_a^b = u(b)v(b) - u(a)v(a).$$

This is the product rule reformulated for integrals. It is most useful, as shown in the example below:

$$\int x^n \log x\,dx = \frac{x^{n+1}}{n+1}\log x - \int \frac{x^{n+1}}{n+1}\frac{1}{x}\,dx$$

$$\int u'v\,dx = u \cdot v - \int uv'dx\,;$$

hence

$$\int x^n \log x\,dx = \frac{x^{n+1}}{n+1}\log x - \frac{1}{n+1}\int x^n dx = \frac{x^{n+1}}{n+1}\log x - \frac{x^{n+1}}{(n+1)^2}.$$

For $n = 0$ we obtain, in particular

$$\int \log x\,dx = x \log x - x.$$

This is quite a surprising result! Directly confronted with the integral, we might not readily have thought of applying our rule by placing $u' = 1$, thus obtaining $u = x$ (while in the case of $\int x^n \log x\,dx$ the idea would have suggested itself more easily). The product rule thus transformed for integrals is called the rule of "integration by parts." An example like the one just treated gives an idea of the wealth of integrations as well as of differentiations which can be performed already with these few rules.

26. FUNCTIONS OF FUNCTIONS

From the product rule we obtain

$$(u^2)' = uu' + u'u = 2uu',$$

$$(u^3)' = u^2u' + (u^2)'u = 3u^2u'.$$

Through complete induction from n to $n + 1$, we can at once prove the formula for all n:

$$(u^n)' = nu^{n-1}u'.$$

For, assuming this formula to be true for some one n, the product rule at once proves it true also for $n + 1$:

$$(u^{n+1})' = (u^nu)' = (u^n)'u + u^nu' = nu^{n-1}u'u + u^nu' = (n + 1)u^nu;$$

that is, if true for n, the rule is true also for $n + 1$. Since it was true for $n = 2$, 3, it is proved true for all n.

In particular, we also find

$$(u^{-1})' = -\frac{1}{u^2}u'.$$

These are special cases of a more general rule:

$$[g(u)]' = g'(u)u'(x).$$

In the two above cases we had, respectively, $g(u) = u^n$ and $g(u) = u^{-1}$.

Leibniz writes this formula in a manner which makes it appear very obvious. The derivative $u'(x)$ is

$$\lim_{x_1 \to x} \frac{u(x_1) - u(x)}{x_1 - x},$$

that is, the limit of the quotient $\Delta u/\Delta x$ of the differences $\Delta u = u(x_1) - u(x)$, and $\Delta x = x_1 - x$, both of which separately approach zero. Now Leibniz conceives of $u'(x)$ likewise as a quotient du/dx of two "differentials"; similarly, he writes for $g'(u)$ the "differential quotient" dy/dx, where $y = g(u)$. Hence, in his notation, the above rule is

$$\frac{dy}{dx} = \frac{dy}{du}\frac{du}{dx}.$$

Now, if these "differentials" were independently defined quantities, we could, in accordance with Leibniz, cancel the du and have the rule completely evident. This is the manner in which Leibniz continued the method of the indivisibles.[40]

We shall of course have to inquire into the exact conditions under which the rule can be rigorously proved.

Let $u(x)$ be differentiable at the value x, and let $g(u)$ be differentiable at the value $u = u(x)$. Let $y = g(u)$. Then we have

$$\frac{y_1 - y}{x_1 - x} = \frac{g(u_1) - g(u)}{x_1 - x} = \frac{g(u_1) - g(u)}{u_1 - u} \cdot \frac{u(x_1) - u(x)}{x_1 - x}.$$

Since $u(x)$ is differentiable at x, u_1 approaches u as x_1 approaches x. Hence the rule follows at once—in fact, hardly less directly than under Leibniz' procedure: instead of the differential, it is the difference which cancels out. But we see that this proof has a fault, although it takes good eyes to see it. We multiplied numerator and denominator by $u_1 - u$. Doing that in a denominator, the mathematician has to be sure that $u_1 - u$ is not equal to zero. Are we sure of that? No! It may very well happen that $u(x_1) = u(x)$! (See Fig. 83.)

But this difficulty is easily taken care of. If it does happen that $u(x_1) = u(x)$, but only once and never again, while $x_1 \rightarrow x$, there is no trouble at all. If, on the other hand, it happens again and again while $x_1 \rightarrow x$ (Fig. 84), then $[u(x_1) - u(x)]/(x_1 - x)$ would again and again assume the value zero. But since, according to hypothesis, $u'(x)$ exists, this limit must equal zero. That is, $u'(x) = 0$ at the value x_1. However, $g(u_1) = g(u)$ for all x_1 for which $u(x_1) = u(x)$, so that $[g(u)]'$ is also equal to zero. Hence the statement becomes $0 = g'(u) \cdot 0$, which is surely true.

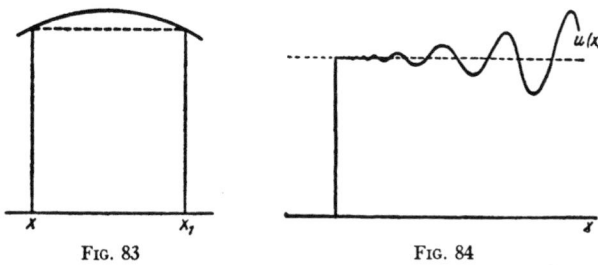

FIG. 83   FIG. 84

Having thus made sure of the validity of the formula, we give an example:

$$y = \log(x^4 + x^2 + 1); \quad \frac{dy}{dx} = \frac{1}{x^4 + x^2 + 1}(4x^3 + 2x).$$

27. TRANSFORMATION OF INTEGRALS

We shall next use this rule for the purpose of obtaining integrals. Beginning with the function of our last example:

1.
$$\int \frac{4x^3 + 2x}{x^4 + x^2 + 1} \, dx.$$

Placing $x^4 + x^2 + 1 = u$, we have $du/dx = 4x^3 + 2x$, or, purely formally, $du = (4x^3 + 2x)dx$. Hence

$$\int \frac{du}{u} = \log u = \log(x^4 + x^2 + 1).$$

2.
$$\int \frac{dx}{x \log x} = \int \frac{1/x}{\log x} \, dx = \int \frac{du}{u} = \log u = \log(\log x),$$

where $\log x = u$, $du = (1/x)dx$.

3. $$\int (ax+b)^n dx = \int t^n \frac{1}{a} \, dt,$$

where $t = ax + b$, $dt = a dx$

$$= \frac{1}{a} \frac{t^{n+1}}{n+1} = \frac{1}{a} \frac{(ax+b)^{n+1}}{n+1}.$$

We thus see the general principle:

$$\int g(u)\,dx = \int g(u)\frac{du}{du/dx} = \int g(u)\frac{dx}{du}\,du.$$

In the integral on the left side u is a function of x. In the right-hand integral we must have a function of u under the integral. This requires that, within the interval ab under consideration, we can express both u as a function of x and x as a function of u. Under what conditions this is possible we shall investigate in the next section.

We still verify our rule for the transformation of integrals through differentiation with respect to x. On the left hand we obtain $g(u)$. Denoting the function on the right side by $G(u)$, we have

$$\frac{dG}{dx} = \frac{dG}{du}\frac{du}{dx} = g(u)\frac{dx}{du}\frac{du}{dx} = g(u).$$

For the definite integral we have

$$\int_a^b g(u)\,dx = \int_\alpha^\beta g(u)\frac{dx}{du}\,du,$$

where $\alpha = u(a)$, $\beta = u(b)$.

This purely formal process is a triumph of Leibniz' formalism of the differential. It shows itself as a very adaptable form for carrying out the computational transformation of the integral. Herein lies the real justification for this formalism. Moreover, there is the advantage that the "indivisibles" are readily visualized; and, as to the differential, it could subsequently be given a meaning by defining dy as $(dy/dx)dx$ or $y'dx$, and regarding dx, the differential of the variable x, as an arbitrarily assigned quantity.

28. THE INVERSE FUNCTION

The inverse of a function is a concept which—perhaps without being called by that name—has often been encountered in elementary mathematics. We use the logarithm tables both to find the logarithm of a number x, $y = \log x$, and to find the number x belonging to a given logarithm y, $x = 10^y$. Strictly speaking, at first $y = 10^x$ is defined, and only afterward $\log y$ is given as the inverse function. That was how Napier and Bürgi had proceeded, except that Napier took $y = e^x$, and obtained from it $x = \log y$. Similarly, to $y = x^2$ belongs the inverse function $x = \sqrt{y}$; in fact, $\sqrt{y}$ is defined as "that number x which squared gives y." Finally, we also encountered arc sin x as the inverse function to sin x, etc.

Using Leibniz' notation, from $y' = dy/dx$, we obtain at once that

$$x' = \frac{dx}{dy} = \frac{1}{dy/dx} = \frac{1}{y'};$$

that is, the rule for differentiating inverse functions. We give a few examples.

1. $\quad y = x^n, \qquad x = \sqrt[n]{y}, \qquad \frac{dy}{dx} = nx^{n-1}, \qquad \frac{dx}{dy} = \frac{1}{nx^{n-1}} = \frac{1}{n} x^{1-n}.$

So far, however, the derivative is still expressed in terms of x, that is, in terms of what is now regarded as the *dependent* variable. We wish to express it, as usual, in terms of the *independent* variable, which is now y. By writing $x = y^{1/n}$, we obtain

$$\frac{dx}{dy} = \frac{1}{n}(y^{1/n})^{1-n} = \frac{1}{n} y^{(1/n - n/n)} = \frac{1}{n} y^{(1/n)-1}.$$

Finally, we can, of course, return to the customary notation, using the symbol y for the dependent, x for the independent, variable, since their naming is an irrelevant matter of form:

$$y = x^{1/q}, \qquad \frac{dy}{dx} = y' = \frac{1}{q} x^{(1/q)-1}.$$

This shows that the rule for $y' = (x^n)' = nx^{n-1}$ holds also if n has the fractional value $1/q$.

2. $\quad y = x^{p/q} = (x^{1/q})^p$

$$y' = p(x^{1/q})^{p-1}(x^{1/q})' = p(x^{1/q})^{p-1} \frac{1}{q} x^{(1/q)-1}$$

$$= \frac{p}{q} x^{(p-1)/q} x^{(1/q)-1} = \frac{p}{q} x^{[(p/q)-(1/q)]+[(1/q)-1]} = \frac{p}{q} x^{[(p/q)-1]};$$

that is, the rule $y' = (x^n)' = nx^{n-1}$ holds also for every fractional value of n.

We could prove the same to be true also for negative fractions; but instead we shall proceed in a more general and comprehensive manner.

3. We shall take x^r for an arbitrary real number r. With this in mind, we should first consider how x^r is defined if r is not a rational number. We saw that Briggs had assumed 10^x as well defined, thinking, for example, of $10^{\sqrt{2}}$ as given by the sequence 10^1, $10^{1.4}$, $10^{1.41}$, $10^{1.414}$, ..., in consideration of the fact that $\sqrt{2} = 1.4142 \ldots$. Briggs, of course, did not investigate whether this sequence is convergent but based the conviction that it was convergent solely on the computational procedure. At this point lies—seen from the standpoint of logic—the interesting break between Napier and his successors. Briggs

simply neglects higher decimals without regard to the consequences. For five-place values he writes, for example, log 2 = 0.30103, and next for log 2^{100} = 100 log 2 he writes 30.10300, without any thought of whether the two zeros at the end is justified. (In this particular example it would indeed happen to be justified since, given to seven places, log 2 = 0.3010300.) In his Preface, how-ever, he himself uses the memorable phrase, *morbus decimalium;* that is, he is aware that what he is doing here is "unhealthy," or, as we would say, inaccurate. This *morbus decimalium* had already shown itself in Ptolemy's sine tables, and, since Briggs, it pervaded all numerical computation, the whole practice of ap-plied mathematics.

In view of all this, we are impressed with the trouble that Napier himself took to keep his statements accurate. He would, for example, find out, and state, whether log 2 is greater than or less than 0.30103, although in the end he could not really go through with it. At any rate, Napier never defined log y in terms of e^x and never used the laws of exponents. Nor did he ever say that a function possessing the principal property is the inverse of a function a^x, that is, the "logarithm to a certain base a," as we say today. For him a^x did not exist. For him log x was defined through the two properties

$$(\log x)' = \frac{1}{x}, \qquad \log 1 = 0,$$

and e^x was the inverse of the function so defined.

We had best follow him in this. Or we define log x—in accordance with Gregorius a Santo Vincentio—as $^x\!\int_1(dt/t)$ and derive from it, with the aid of the fundamental theorem, the above two properties, as well as the formula log ab = log a + log b, as was done previously. That is doubtless still the best way to define e^x; $e^{x+y} = e^x e^y$ follows from it.

And what is then a^x? We define it as $a^x = e^{x \log a}$. Then we have

$$a^{x+y} = e^{(x+y) \log a} = e^{x \log a + y \log a} = e^{x \log a} e^{y \log a} = a^x a^y ;$$

that is, the rules of exponents follow at once for irrational exponents. After this preparation we can tackle the differentiation of $y = a^x$.

$$\underline{y = a^x = e^{x \log a} = e^u} ; \quad u = x \log a ;$$

hence

$$y' = (e^u)' \frac{du}{dx} = e^u \log a = \underline{a^x \log a} .$$

Similarly, we can now differentiate $y = x^a$:

$$\underline{y = x^a = e^{a \log x} = e^u}, \qquad u = a \log x, \qquad u' = \frac{a}{x}$$

$$y' = (e^u)' \frac{du}{dx} = e^u \frac{a}{x} = x^a \frac{a}{x} = \underline{a x^{a-1}},$$

that is, the rule $(x^v)' = vx^{v-1}$ holds for all real numbers v. Correspondingly, we obtain

$$\int x^v dv = \frac{x^{v+1}}{v+1} \quad \text{for } v \neq -1.$$

So far we have only demonstrated the rule for differentiating inverse functions through examples—and wisely so. For here, too, we shall first have to straighten out things, and define precisely *what an inverse function is* before we can talk about its derivative; the latter question should then cause no particular difficulty.

We said that "$\sqrt{y}$ is the number x whose square is y"; in that way we explained $x = \sqrt{y}$ as the inverse to the function $y = x^2$. We said "*the* number." But in fact there are two numbers; for example, in the case of $y = 9$, the two numbers $+3$ and -3, and similarly for every $y > 0$, namely, $+y^{1/2}$ (which is

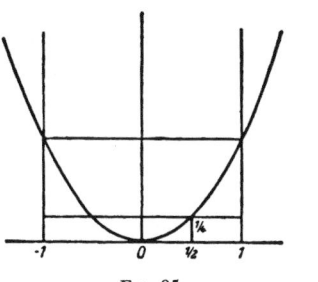

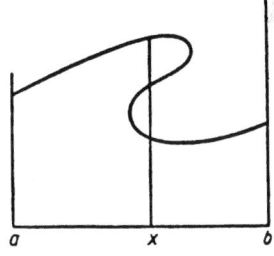

FIG. 85   FIG. 86

now neatly defined as $e^{(1/2)\,\log x)}$ and $-y^{1/2}$. For $y < 0$, on the other hand, *the* number $x = \sqrt{y}$ does not exist, since the square of every positive number is positive, and the square of every negative number is positive, and $0^2 = 0$, which means that we never obtain a negative number by squaring any number. The phrase "the number x" thus proves to be thoroughly misleading. In practical applications we can often get by with it. If, for example, x represents the volume of sound, it cannot be negative, wherefore it is clear that only the positive value is meaningful. But in exact mathematics this manner of defining is inadmissible. Throughout the eighteenth century, mathematicians proceeded in this imprecise way. When things became more complicated, however, they got into trouble and had to start cleaning up, which was done very thoroughly in the course of the nineteenth century.

Geometrical presentation is very helpful to show what is needed, for there is no conflict between logic and presentation as long as we do not assign to the latter a role which it cannot play. We should first clearly understand that, while every function $y = f(x)$ can be geometrically represented (Fig. 85), not every curve arbitrarily drawn is the geometrical representation of a function (Fig. 86).

For we must keep in mind that a function is a rule which assigns to every value x in an interval $a \leq x \leq b$ *one*, and *only one*, value $y = f(x)$. Consequently, every vertical constructed in the interval ab intersects the graph of a function once and once only. Curves which are intersected more than once by some verticals in the interval, or not at all, cannot be representations of functions (Figs. 86–87).

A curve is the geometrical representation of a function f(x) *in the interval* ab *if and only if it is intersected by every vertical in the interval once and once only.*

For $y = x^2$ everything is all right. Here the parabola satisfies these conditions not only in the interval $-1 \leq x \leq 1$ but in every interval extending no matter how far to the right or to the left, as we might say, in the interval $-\infty < x < +\infty$. (To designate the ends of an interval, which we do not use for computation, this dangerous symbol may be used safely, since it is perfectly clear what is meant by it.)

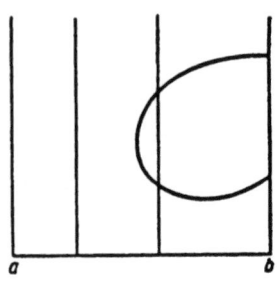

Fig. 87

If next we ask, "For which values of x is $x^2 = \frac{1}{4}$?" we have to consider the horizontal line drawn at the distance $\frac{1}{4}$ from the x-axis and find its intersections with the parabola. The abscissas of the two intersections are $x = \frac{1}{2}$, $x = -\frac{1}{2}$. Altogether, we find the x belonging to any assigned y by constructing the horizontal at the distance y, or of ordinate y. For $y > 0$ the horizontal cuts the parabola in two points; for $y < 0$ it does not cut the parabola at all—that is all clearly apparent to the eye.

This example shows that, to get at the inverse function, we have to consider the horizontals instead of the verticals in the graph. Where, as was the case here, the horizontals intersect the curve more than once, or not at all, we do not obtain a clear-cut concept of the inverse of a function. We can avoid this difficulty if we remember that by its very nature a function is defined for a given interval. In our example we neglected to consider the interval, because $y = x^2$ was defined in every interval. We have to return to a definite interval and restrict its limits in such a way that the horizontal intersects the curve in this interval exactly once. If in the case of $y = x^2$ we choose the interval $0 \leq x \leq a$, where a can be arbitrarily large, then every vertical from zero to a meets the curve exactly once, and every horizontal from zero to a meets the curve likewise exactly once (Fig. 88).

What is the general condition for this to happen? In Barrow's proof of the fundamental theorem we had formulated the condition for $f(x)$ to be *monotonic* and *continuous* in the whole interval. We shall see that this is the condition which applies here too. The monotonicity guarantees that no horizontal (with an ordinate ξ) cuts the curve more than once (Fig. 89). For, if it should happen anywhere that $f(x_1) = f(x_2) = \xi$, this would mean that $f(x)$ would not increase from x_1 to x_2, but remain unchanged. By "monotonic" we mean, how-

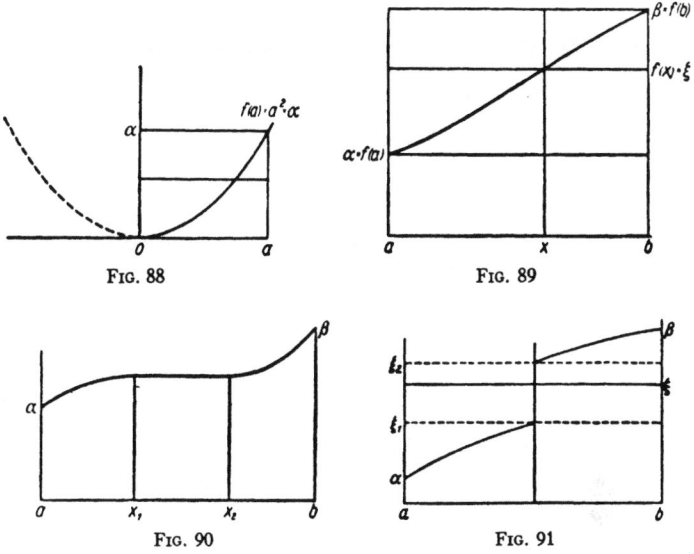

FIG. 88   FIG. 89

FIG. 90   FIG. 91

ever, a "properly monotonic" function which never stays constant but for which $f(x_1) < f(x_2)$, not $f(x_1) \leq f(x_2)$, for $x_1 < x_2$, which was quite admissible in the definition of the definite integral (Fig. 90).

The continuity of $f(x)$, on the other hand, guarantees that every horizontal with ordinate ξ from the interval $a = f(a) \leq \xi \leq f(b) = \beta$ really does intersect the curve. If $f(x)$ could take a jump, as in Figure 91, the ordinates ξ_1 to ξ_2 would be "cut out," even though $f(x)$ would still have to be regarded as properly monotonic; but it would not be continuous. That $f(x)$, if continuous everywhere between a and b, actually assumes *all* values between a and β seems very obvious indeed. In fact, this should, of course, be proved to follow from the definition of continuity given above. This is the second time that we encounter a very evident fact which yet calls for a mathematical proof. The first time it was a fact that a movable point which rests at every moment is at rest altogether. We shall again postpone the proof until later and for the present let it go with the intuitively clear evidence. We may now summarize:

If a function ξ = f(x) *is defined in an interval* a ≤ x ≤ b, *and if it is continuous and monotonic, then it has an inverse function* x = φ(ξ), *which is (unambiguously) defined in the interval* φ = f(a) ≤ ξ ≤ β = f(b), *and which determines for every* ξ *that* x = φ(ξ) *for which* f(x) = ξ, *or* f[φ(ξ)] = ξ. *Then for every* x *in the interval* a ≤ x ≤ b *the relation* x = φ[f(x)] *holds.*

We may now change our notation and write the inverse function as $y = φ(x)$, not $x = φ(ξ)$, so as not to have the interval $αβ$ vertical, to be read off on the side, and make it necessary to glance along the horizontals. We rather shall

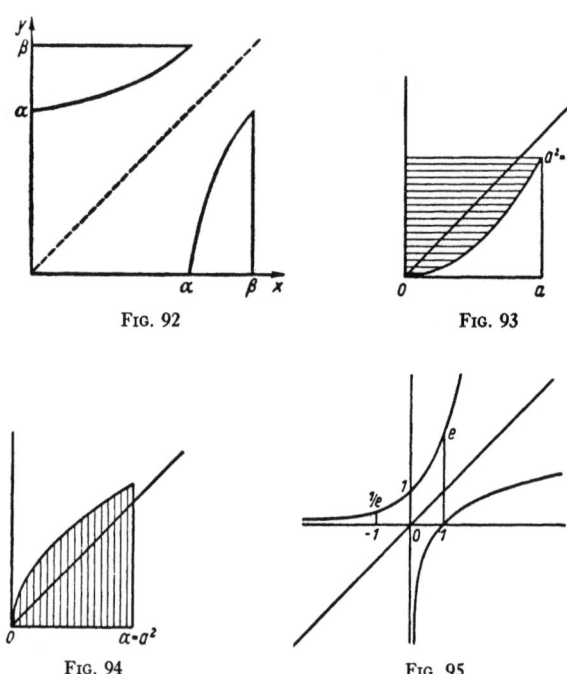

FIG. 92   FIG. 93

FIG. 94   FIG. 95

make a mirror reflection of the whole figure on the bisector of the angle between the two co-ordinate axes and thus obtain the graph of the inverse function in the customary upright orientation (Fig. 92).

For the parabola we would then use Figure 94 instead of Figure 93. That gives the graph of $y = \sqrt{x}$. Since a can be arbitrarily large, we can say that this function $\sqrt{x}$ is well defined in the whole interval $0 ≤ x < ∞$.

For $y = e^x$ the definition is one-valued for all real x just as in the case of $y = x^2$. Further, $e^{-x} = 1/e^x$, $e^x ≠ 0$ for all values of x, and $e^0 = 1$. The function is throughout properly monotonic and continuous; it increases from 0 to ∞ (of course e^x is never *equal* to zero!). Hence the inverse function $y = \log x$ is defined in the interval $0 < x < ∞$, with $\log 1 = 0$ (Fig. 95).

29. TRIGONOMETRIC FUNCTIONS

We start with the differentiation of the sine function. According to the addition theorem, we have

$$\frac{\sin x_1 - \sin x}{x_1 - x}$$

$$= \frac{\sin\{[(x_1+x)/2] + [(x_1-x)/2]\} - \sin\{[(x_1+x)/2] - [(x_1-x)/2]\}}{x_1 - x}$$

$$= \frac{2 \cos [(x_1+x)/2] \sin [(x_1-x)/2]}{x_1 - x}$$

$$= \frac{\cos\left(\frac{x_1+x}{2}\right) \sin [(x_1-x)/2]}{(x_1-x)/2}.$$

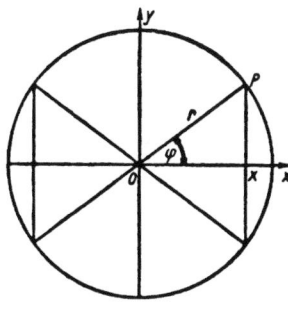

FIG. 96

For $x_1 \to x$, $\cos (x_1 + x)/2$ approaches $\cos x$, if we presuppose the continuity of $\cos x$, a point about which we shall say more a little later. More difficult is the other factor. Here both numerator and denominator approach zero; the factor has the form

$$\lim_{h \to 0} \frac{\sin h}{h},$$

where

$$h = [(x_1 - x)/2].$$

So we need two things: (1) $\lim_{h \to 0}$ (sin h/h), and (2) the continuity of the function $\cos x$. To answer the first, we have to go back to the definition of sin x. There is nothing difficult about that. Sin φ (Fig. 96) is the perpendicular PX, or, rather, its numerical measure in terms of the radius r of the circle which is taken as unity. It is counted positive if P lies above OX; negative, if P lies below OX. Hence sin φ is positive if φ increases from zero to $2R$, where R is a right angle, negative when φ increases from $2R$ to $4R$, the full angle; $4R$ is indistinguishable from zero.

The next question is in what units we measure angles. The Babylonians had a sexagesimal system; they measured the right angle, which was their unit, in this system (that is, they subdivided it into 60, then into 60^2, etc., parts). If we proceeded in that way with our decimal system, we would use $0.1R$, $0.2R$, etc., as our decimal subunits. But, although adopting the Babylonian subdivisions, we have replaced the sexagesimal by the decimal number system. That has brought about a disharmony which is very bothersome for astronomers. But they cannot do much about it, since to change it would mean to cast aside all those valuable trigonometric tables which we have and to replace them by new ones, which would involve enormous labor and expense as well as endless computational errors.

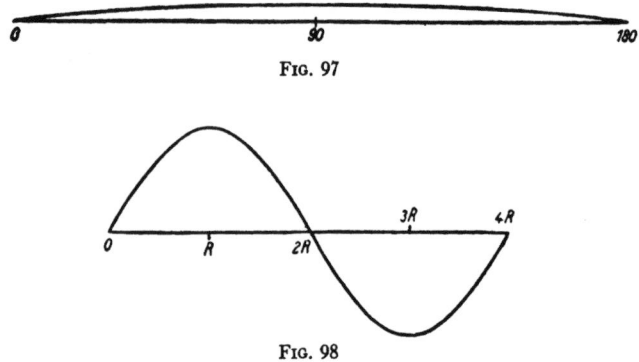

FIG. 97

FIG. 98

Moreover, it has become customary to take not R as the unit but $\frac{1}{90}$ of it, called $1°$. This is the unit in which angles are measured. How impractical it is we see when trying to draw the graph of the function sin x. The variable x has to run from $0°$ to $360°$, while $y = \sin x$ never rises or falls by more than 1 unit. That would give a very impractical, long-drawn-out graph. The one shown in Figure 97 does not represent the true dimensions by any means.

It might be better to plot R as unity. The graph then becomes the one shown in Figure 98. But from a mathematical standpoint this would not be a good idea either. We are here in a situation similar to that which arose in connection with logarithms. For purposes of numerical computations the base 10 seemed to recommend itself; but then y' would not be simply $1/x$ but c/x, where c is an inconvenient factor which would trail through all formulas.

We remember from the measurement of the circle that (Fig. 99) the chord PP' is less than the arc $\overparen{PEP'}$ and the arc $\overparen{EPR}$ less than the broken line EQR, for a convex curve is always longer than any other completely inclosed by it (cf. Sec. 17).

Now $PX = \sin \varphi$, $QE = \tan \varphi$; hence $PP' = 2 \sin \varphi$, $EQ + QR = 2 \tan \varphi$. Hence

$$2 \sin \varphi < \overparen{PEP}, \qquad \overparen{EPR} < 2 \tan \varphi,$$

or

$$\overparen{EP} < \frac{\sin \varphi}{\cos \varphi}, \qquad \sin \varphi < \overparen{EP},$$

Therefore,

$$\overparen{EP} \cos \varphi < \sin \varphi < \overparen{EP},$$

or

$$\cos \varphi < \frac{\sin \varphi}{\overparen{EP}} < 1.$$

In all this it is assumed that the radius r of the circle is the unit of length, that is, $r = 1$.

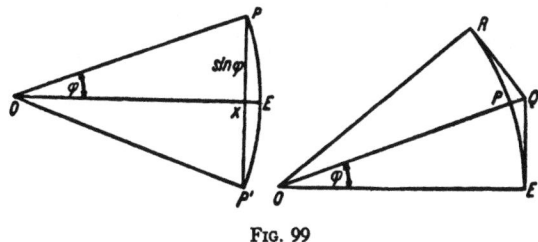

FIG. 99

Just as logarithm tables for different bases differ from each other only by a constant factor, so all possible choices of the angular unit differ only by a factor. The only question is which angle to choose as unity. If now we decide to choose the length of the arc $\overparen{EP}$ as the numerical measure of the angle, then $\overparen{EP} = \varphi$, and consequently the above inequality becomes

$$\cos \varphi < \frac{\sin \varphi}{\varphi} < 1.$$

When φ approaches zero, $\cos \varphi$ approaches 1; hence for the middle term we obtain

$$\lim_{\varphi \to 0} \frac{\sin \varphi}{\varphi} = 1.$$

For this angular unit the right angle R is measured by the number $\pi/2$, the straight angle by π; $1°$ by $\pi/180$, etc. If instead we were to choose R as unit angle, it would have the measure 1 rather than $\pi/2$, and consequently the measures of all angles would be multiplied by $2/\pi$. In that case we would find

$$\lim_{\varphi \to 0} \frac{\sin \varphi}{\varphi} = \frac{\pi}{2},$$

and hence the number π would have to be dragged through all the formulas which we are going to derive next. It was to avoid this nuisance that we have defined the angular unit as we did.

Turning to the other trigonometric functions, cos x is defined as sin $(\pi/2 - x)$ tan x as sin x/cos x. There also hold the addition theorems:

$$\sin (x + y) = \sin x \cos y + \cos x \sin y ,$$

$$\cos (x + y) = \cos x \cos y - \sin x \sin y .$$

From $\lim_{x \to 0} \sin x = 0$, the definition for cos x shows at once $\lim_{x \to 0} \cos x = 1$. Because of this the addition theorems prove the continuity of the two functions. For example, in cos $(x + y)$, if $y \to 0$, the right side approaches cos x.

Now we have everything we need. Returning to the difference quotient which we had formed, we find that, as x_1 approaches x,

$$\cos \frac{x_1 + x}{2} \to \cos x , \qquad \frac{\sin [(x_1 - x)/2)]}{(x_1 - x)/2} \to 1 ;$$

hence $(\sin x)' = \cos x$.

The function sin x is defined for all values x; it is continuous and differentiable, and $(\sin x)' = \cos x$.

From this result we derive at once:

$$(\cos x)' = \left[\sin \left(\frac{\pi}{2} - x \right) \right]' = \cos \left(\frac{\pi}{2} - x \right) \left(\frac{\pi}{2} - x \right)' = -\cos \left(\frac{\pi}{2} - x \right) = -\sin x$$

$$(\tan x)' = \left(\frac{\sin x}{\cos x} \right)' = \frac{\cos^2 x + \sin^2 x}{\cos^2 x} = \frac{1}{\cos^2 x}$$

$$(\cot x)' = \left[\tan \left(\frac{\pi}{2} - x \right) \right]' = -\frac{1}{\sin^2 x} .$$

30. INVERSE TRIGONOMETRIC FUNCTIONS

Purely formally, we obtain:

$$y = \sin x , \qquad x = \text{arc sin } y , \qquad \frac{dx}{dy} = \frac{1}{\cos x} = \frac{1}{\sqrt{1 - y^2}} :$$

$$(\text{arc sin } x)' = \frac{1}{\sqrt{1 - x^2}} ,$$

$$y = \cos x , \qquad x = \text{arc cos } y = \frac{\pi}{2} - \text{arc sin } y$$

$$\frac{dx}{dy} = - \frac{1}{\sqrt{1 - y^2}} : \qquad (\text{arc cos } x)' = - \frac{1}{\sqrt{1 - x^2}}$$

$$\left[y = \sin \left(\frac{\pi}{2} - x \right) , \quad \frac{\pi}{2} - x = \text{arc sin } y \right] ,$$

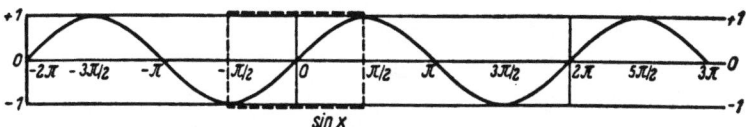

Fig. 100

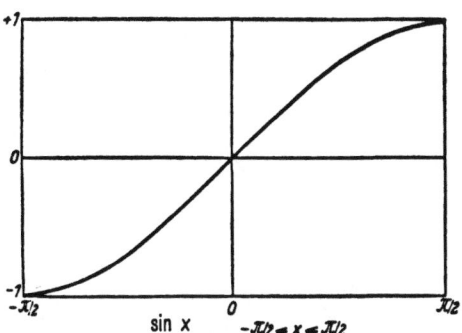

sin x   −π/2 ≤ x ≤ π/2

Fig. 101

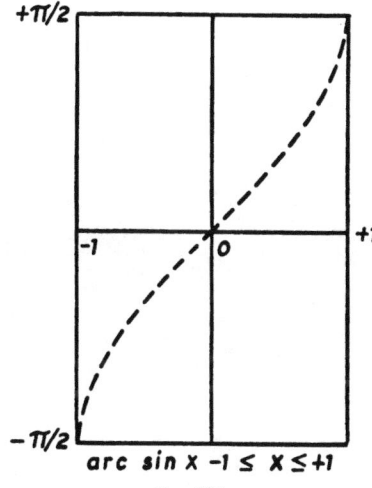

arc sin x −1 ≤ x ≤ +1

Fig. 102

$$y = \tan x, \qquad x = \operatorname{arc\ tan} y,$$

$$\frac{dy}{dx} = \frac{1}{\cos^2 x} = \frac{\cos^2 x + \sin^2 x}{\cos^2 x} = 1 + \tan^2 x = 1 + y^2,$$

$$\frac{dx}{dy} = \frac{1}{1+y^2}: \qquad (\operatorname{arc\ tan} x)' = \frac{1}{1+x^2}.$$

Now for exact definitions. Figure 100 shows the graph of sin x, x being given in arc measurement. Each monotonic part of this function gives rise to an inverse function. Even when aiming at a greatest possible connected part, we

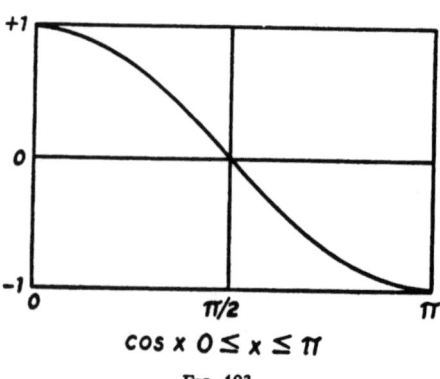

cos x $0 \leqq x \leqq \pi$

FIG. 103

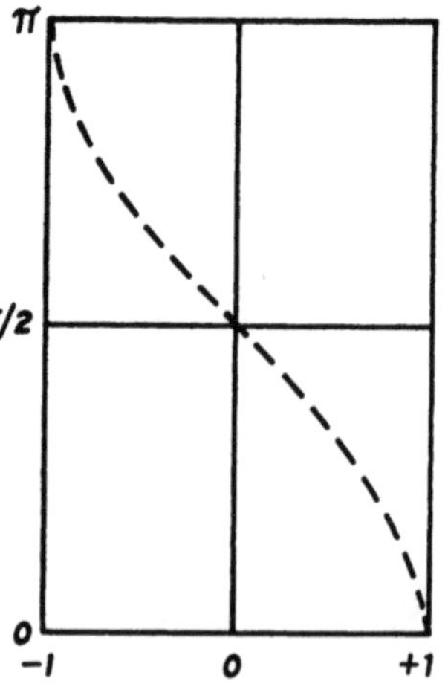

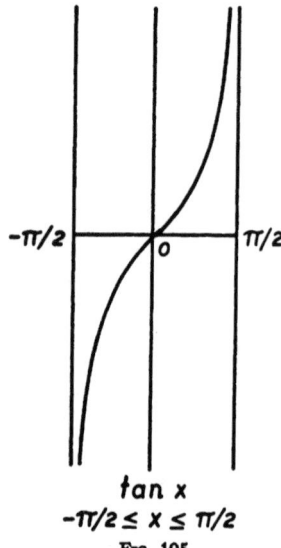

arc cos x $-1 \leqq x \leqq 1$

FIG. 104

tan x
$-\pi/2 \leqq x \leqq \pi/2$

FIG. 105

would still have the choice between infinitely many inverse functions. Mathematicians have agreed to use the interval $-\pi/2 \le x < \pi/2$.

In the adjacent group of figures (Figs. 101-6) the so-called principal branch of each of the inverse trigonometric functions is shown. The other "branches" for arc tan x are obtained from the principal one by merely adding integer multiples of π: . . . $-2\pi +$ arc tan x, $-\pi +$ arc tan x, arc tan x, $\pi +$ arc tan x, $2\pi +$ arc tan x, ...

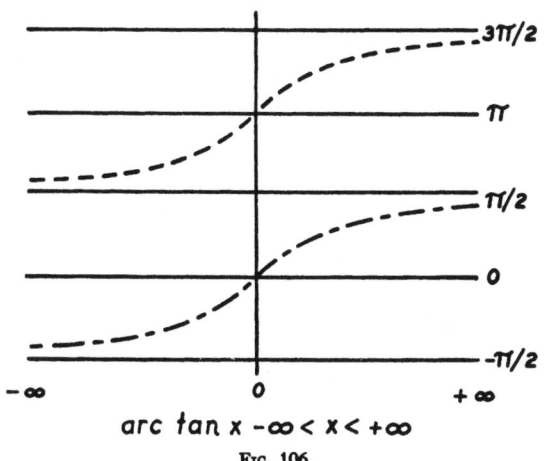

$$\text{arc tan } x \quad -\infty < x < +\infty$$

FIG. 106

31. FUNCTIONS OF SEVERAL FUNCTIONS

If we have to differentiate a function such as

$$f(x) = \frac{\cos^3 x + \sin^3 x}{\cos^3 x - \sin^3 x},$$

it would be convenient to place $\cos x = u$, $\sin x = v$, such that

$$f(x) = \frac{u^3 + v^3}{u^3 - v^3} = F(u, v),$$

where u and v are each functions of x. How do we now find $f'(x)$? Placing $u(x_1) = u_1$ and $v(x_1) = v_1$, we have

$$\frac{f(x_1) - f(x)}{x_1 - x} = \frac{F(u_1, v_1) - F(u, v)}{x_1 - x} = \frac{F(u_1, v_1) - F(u, v_1)}{u_1 - u} \frac{u_1 - u}{x_1 - x}$$

$$+ \frac{F(u, v_1) - F(u, v)}{v_1 - v} \frac{v_1 - v}{x_1 - x}.$$

Now, as x_1 approaches x, $(u_1 - u)/(x_1 - x)$ approaches $u'(x)$ and $(v_1 - v)/(x_1 - x)$ approaches $v'(x)$. Therefore

$$\frac{F(u, v_1) - F(u, v)}{v_1 - v}$$

approaches the derivative of $F(u, v)$ with respect to v, in which u plays the role of a constant. We designate this derivative as $F_2'(u, v)$. More difficult is the question as to what becomes of

$$\frac{F(u_1, v_1) - F(u, v_1)}{u_1 - u}.$$

For, as x_1 approaches x, v_1 does not remain constant as did u but approaches v. Now it is very plausible that the quotient in question does approach the derivative of $F(u, v)$ with respect to u, with v being kept constant, which we designate by $F_1'(u, v)$; but this is a point which we shall have to consider more closely later on (so far the third point postponed for later discussion). Accepting this result, we then have

$$F'(x) = F_1'(u, v) u'(x) + F_2'(u, v) v'(x) = \frac{\partial F}{\partial x} \frac{du}{dx} + \frac{\partial F}{\partial v} \frac{dv}{dx}.$$

The second expression gives a new way of writing the derivatives F_1' and F_2', where ∂ indicates that only the one of the variables u and v varies while the other stays constant. (As in Section 24, we should really devote an additional discussion to the case of vanishing denominators $u_1 - u$ and $v_1 - v$.)

The function of our example thus can be differentiated as follows:

$$\left(\frac{\cos^3 x + \sin^3 x}{\cos^3 x - \sin^3 x}\right)' = \frac{\partial}{\partial u}\left(\frac{u^3 + v^3}{u^3 + v^3}\right)(-\sin x) + \frac{\partial}{\partial v}\left(\frac{u^3 + v^3}{u^3 - v^3}\right)\cos x$$

$$= -\frac{3u^2(u^3 - v^3) - 3u^2(u^3 + v^3)}{(u^3 - v^3)^2}\sin x$$

$$+ \frac{3v^2(u^3 - v^3) + 3v^2(u^3 + v^3)}{(u^3 - v^3)^2}\cos x$$

$$= \frac{6u^2 v^3}{(u^3 - v^3)^2}\sin x + \frac{6u^3 v^2}{(u^3 - v^3)^2}\cos x$$

$$= \frac{6u^2 v^2(u^2 + v^2)}{(u^3 - v^3)^2} = \frac{6\cos^2 x \sin^2 x}{(\cos^3 x - \sin^3 x)^2}.$$

For another example let us differentiate $f(x) = x^x$. As to x^x, we may treat it as $e^{x \log x}$. From this we find

$$f'(x) = e^{x \log x}(x \log x)' = x^x(\log x + 1).$$

Or else we can treat x^x as

$$y = x^x = u^v(= e^{v \log u}),$$

where $u = x$ and $v = x$. From this we obtain

$$y' = \frac{\partial}{\partial u}(u^v)\frac{du}{dx} + \frac{\partial}{\partial v}(u^v)\frac{dv}{dx} = v u^{v-1} + u^v \log u$$

$$= x x^{x-1} + x^x \log x = x^x + x^x \log x = x^x(1 + \log x).$$

32. INTEGRATION OF RATIONAL FUNCTIONS

We are already able to find the indefinite integrals for a great many functions by inverting known formulas for differentiation. But these are what we might call accidental results. Let us now proceed in a more systematic manner.

We quickly learned to differentiate and to integrate polynomial functions thus:

$$\int (a_0 + a_1 x + \ldots + a_n x^n)\, dx = a_0 x + \frac{a_1}{2} x^2 + \frac{a_2}{3} x^3 + \ldots + \frac{a_n}{n+1} x^{n+1}.$$

The quotient rule also permitted us to differentiate all rational functions. Can we also integrate all rational functions? We shall give below a summary of various cases of this class of functions.

Since $(ax+b)^{n+1}$ has the derivative $(n+1)(ax+b)^n a$, we have

$$\int (ax+b)^n dx = \frac{1}{a}\frac{1}{n+1}(ax+b)^{n+1}.$$

This holds also for negative n, provided $n + 1 \neq 0$, or $n \neq -1$:

Further

$$\int \frac{dx}{(ax+b)^m} = \frac{1}{a}\frac{1}{1-m}\frac{1}{(ax+b)^{m-1}}, \qquad m \neq 1.$$

Also

$$\int \frac{dx}{ax+b} = \frac{1}{a}\log(ax+b).$$

or, if

$$\int \frac{f(x)}{x-a}\, dx = \int \frac{f(x)-f(a)}{x-a}\, dx + \int \frac{f(a)}{x-a}\, dx,$$

$$f(x) = a_0 + a_1 x + \ldots + a_n x^n,$$

$$f(a) = a_0 + a_1 a + \ldots + a_n a^n,$$

we have

$$\int \frac{f(x)}{x-a}\, dx = \int [a_1 + a_2(x+a) + \ldots a_n(x^{n-1} + \ldots a^{n-1})]\, dx$$
$$+ f(a)\log(x-a).$$

The first integral is that of a polynomial which can be computed; hence the whole integral can be found. Finally,

$$\int \frac{f(x)}{ax+b}\, dx = \int \frac{1}{a}\frac{f(x)}{x+(b/a)}\, dx = \int \frac{1}{a}\frac{f(x)}{x-a}\, dx,$$

where $a = -(b/a)$. Hence this integral is reduced to the previous one.

Next we come to

$$J_v = \int \frac{f(x)}{(x-a)^v}\, dx = \int \frac{f(x)-f(a)}{(x-a)^v}\, dx + f(a)\int \frac{dx}{(x-a)^v}.$$

We already know the second integral. The first we write in the form

$$\int \frac{f(x)-f(a)}{x-a}\frac{dx}{(x-a)^{v-1}}.$$

Again this quotient is fractional in appearance only; actually it is a polynomial $A_1 + A_1 x + \ldots A_{n-1} x^{n-1}$. Hence the first integral has the form

$$\int \frac{g(x)}{(x-a)^{v-1}}\, dx\,;$$

that is, it is of the same form as J_v, except that v has been replaced by $v - 1$! Thus J_2 is reduced to J_1, J_3 to J_2, etc. Hence we see that, in principle, we can integrate J_2. As before we can then also obtain $\int [f(x)/(ax + b)^v]dx$.

We have thus quickly obtained a rather general result. Nevertheless, integration is not as easy as differentiation. We have by no means shown how to integrate any rational function. As regards the numerator function, it was indeed quite general, but not the denominator function. For not every polynomial of degree n is the nth power of a linear function $(ax + b)^n = a^n x^n + n a^{n-1} b x^{n-1} + \ldots + b^n$. Already $A x^2 + B x + C$ is not always of the form $(ax + b)^2 = a^2 x^2 + 2abx + b^2$; for this would mean $B^2 = 4a^2 b^2$, hence $B^2 - 4AC = 0$, a condition which, in general, is not satisfied.

But we do know already some integrals with a second-degree function in the denominator besides $\int [dx/(ax + b)^2]$, namely:

$$\int \frac{dx}{x^2 + 1} = \text{arc tan } x$$

and

$$\int \frac{dx}{(x-a)(x-\beta)}$$

$$= \int \left(\frac{1}{x-a} - \frac{1}{x-\beta} \right) \frac{1}{a-\beta}\, dx$$

$$= \int \frac{1}{a-\beta} \frac{dx}{x-a} - \int \frac{1}{a-\beta} \frac{dx}{x-\beta}$$

$$= \frac{1}{a-\beta} \log(x-a) - \frac{1}{a-\beta} \log(x-\beta) = \frac{1}{a-\beta} \log \frac{x-a}{x-\beta}.$$

Both these results can be generalized. In the case of $\int [dx/(A x^2 + B x + C)]$ we can write—using a little trick familiar from the solution of quadratic equations—

$$A x^2 + B x + C = A \left(x + \frac{B}{2A} \right)^2 + \left(C - \frac{B^2}{4A} \right)$$

$$= A \left[\left(x + \frac{B}{2A} \right)^2 + \frac{4AC - B^2}{4A^2} \right]$$

$$= A \left[\left(x + \frac{B}{2A} \right)^2 - \frac{B^2 - 4AC}{4A^2} \right].$$

Now we consider three cases.

1. If $B^2 - 4AC = 0$, we have $Ax^2 + Bx + C = A[x + (B/2A)]^2$, the square of a first-degree polynomial. We already know how to integrate that.

2. If $B^2 - 4AC < 0$, then $4AC - B^2 > 0$, and hence $(4AC - B^2)/4A^2 > 0$. Let $u = x + (B/2A)$, then

$$\int \frac{dx}{A x^2 + Bx + C} = \frac{1}{A} \int \frac{du}{u^2 + \delta^2},$$

where $\delta^2 = (4AC - B^2)/4A^2$, or, placing $u = \delta v$,

$$\int \frac{dx}{A x^2 + Bx + C} = \frac{1}{A} \frac{1}{\delta^2} \int \frac{\delta\, dv}{v^2 + 1} = \frac{1}{A} \frac{1}{\delta} \text{ arc tan } v.$$

Thus the second case is also solved.

3. If $B^2 - 4AC > 0$, we place $\delta^2 = (B^2 - 4AC)/4A^2$, and $x + (B/2A) = u$. Then $Ax^2 + Bx + C = A(u^2 - \delta^2) = A(u + \delta)(u - \delta)$; that is the form treated above, with $\alpha = -\delta$ and $\beta = \delta$. Thus we can integrate the form

$$\int \frac{dx}{A x^2 + Bx + C}$$

under all circumstances.

Through a somewhat lengthy procedure

$$\int \frac{f(x)\, dx}{(A x^2 + Bx + C)^v}$$

can be reduced to the above. Again, that does not yet give us the general case by any means. We just noticed that when the denominator is a second-degree polynomial, or a power of it, the problem involves the solution of quadratic equations. We may suspect that in the case of a third-degree polynomial the solution of cubic equations will be involved and that, accordingly, our integration problem becomes more and more difficult. In quite a different context we shall see how Leibniz—overtaking Newton—had the right idea about this problem, without however carrying it out. All that lies along a different line of approach. The final result will be that every rational function can be integrated.

33. INTEGRATION OF TRIGONOMETRIC EXPRESSIONS

If we assume the above result in its full generality, we can derive from it a great many other integrals. First, we can then obtain integrals such as, for example,

$$\int \frac{\sin^6 x - \cos^4 x}{3 \sin^4 x - 2 \cos^2 x}\, dx,$$

or, in general, every integral of the form

$$R(\cos x, \sin x)dx,$$

where R is any rational function $R(u, v)$, with $u = \cos x$ and $v = \sin x$.

The proof rests on a trick going back to no less a mathematician than Weierstrass. He places $t = \tan(x/2)$, that is, $x = 2$ arc tan t; hence $dx/dt = 2/(1 + t^2)$. Then we obtain

$$\frac{2t}{1+t^2} = \frac{2 \tan(x/2)}{1+\tan^2(x/2)} = \frac{2\,\dfrac{\sin(x/2)}{\cos(x/2)}}{1+\dfrac{\sin^2(x/2)}{\cos^2(x/2)}} = \frac{2\sin(x/2)\cos(x/2)}{\cos^2(x/2)+\sin^2(x/2)} = \sin x$$

$$\frac{1-t^2}{1+t^2} = \frac{1-\dfrac{\sin^2(x/2)}{\cos^2(x/2)}}{1+\dfrac{\sin^2(x/2)}{\cos^2(x/2)}} = \frac{\cos^2(x/2)-\sin^2(x/2)}{\cos^2(x/2)+\sin^2(x/2)} = \cos x \,.$$

Hence

$$\int R(\cos x, \sin x)\, dx = \int R\left(\frac{1-t^2}{1+t^2}, \frac{2t}{1+t^2}\right)\frac{2}{1+t^2}\, dt,$$

and the problem is reduced to an integral of rational functions. *Example:*

$$\int \frac{dx}{\sin x} = \int \frac{1+t^2}{2t}\frac{2}{1+t^2}\, dt = \int \frac{dt}{t} = \log t = \log \tan \frac{x}{2}\,.$$

On the other hand, an integral like $\int (x/\sin x)dx$ cannot be treated in this way. It is not of the type $R(\cos x, \sin x)$, because it contains also x.

34. INTEGRATION OF EXPRESSIONS INVOLVING RADICALS

We turn to yet another kind of integral. The area of the circle, or rather of its upper half, whose equation is $x^2 + y^2 = r^2$, or $y = \sqrt{r^2 - x^2}$, if given by the definite integral $\int_{-r}^{r}\sqrt{r^2 - x^2}dx$ (Fig. 107). It is to be clearly understood that $\sqrt{}$ signifies that the positive root is to be taken. We thus have to find the indefinite integral $\int \sqrt{r^2 - x^2}dx$. Let $x = r \cos \varphi$; then $dx = -r \sin \varphi d\varphi$, and $\sqrt{r^2 - x^2} = r \sin \varphi$. Hence

$$\int \sqrt{r^2 - x^2}dx = -\int r^2 \sin^2 \varphi d\varphi = -r^2 \int \sin^2 \varphi d\varphi \,.$$

This is an integral of the type discussed in the preceding section.

By virtue of $\cos 2\varphi = 1 - 2 \sin^2 \varphi$, or $\sin^2 \varphi = \frac{1}{2} - \frac{1}{2}\cos 2\varphi$, we obtain

$$\int \sin^2 \varphi d\varphi = \tfrac{1}{2}\varphi - \tfrac{1}{4}\sin 2\varphi \,.$$

Hence

$$\int_{-r}^{r}\sqrt{r^2 - x^2}dx = -r^2\int_{\pi}^{0}\sin^2 \varphi d\varphi = -r^2[\tfrac{1}{2}\varphi - \tfrac{1}{4}\sin 2\varphi]_{\pi}^{0}$$

$$= +r^2[\tfrac{1}{2}\varphi - \tfrac{1}{4}\sin 2\varphi]_{0}^{\pi} = \frac{\pi}{2} r^2 \,.$$

This is the area of the semicircle; that of the full circle is therefore πr^2.

That is no great surprise. After all, π was defined as the area of a circle of radius 1. Our result is, therefore, merely a verification of our method.

For the area of an ellipse (Fig. 108) we proceed analogously: the equation of the ellipse is

$$\frac{x^2}{a^2}+\frac{y^2}{b^2}=1\ ;\qquad y=b\sqrt{1-\frac{x^2}{a^2}}=\frac{b}{a}\sqrt{a^2-x^2}.$$

Hence the area of the half-ellipse

$$\int_{-a}^{a}\frac{b}{a}\sqrt{a^2-x^2}dx=\frac{b}{a}\int_{-a}^{a}\sqrt{a^2-x^2}dx\ .$$

But ${}^a\!\int_{-a}\sqrt{a^2-x^2}dx$ is the same as ${}^r\!\int_{-r}\sqrt{r^2-x^2}dx$, which was just found to be $\frac{1}{2}r^2\pi$. Hence the area of the ellipse is $(b/a)a^2\pi=ab\pi$.

This example brings to our attention integrals of the form $\int R(x,\sqrt{a^2-x^2}dx)$. They can readily be integrated. Placing again $x=a\cos\varphi$, we obtain $\sqrt{a^2-x^2}=a\sin\varphi,\,dx=-a\sin\varphi d\varphi$. Hence $\int R(x,\sqrt{a^2-x^2})dx=\int R^*(\cos\varphi,$

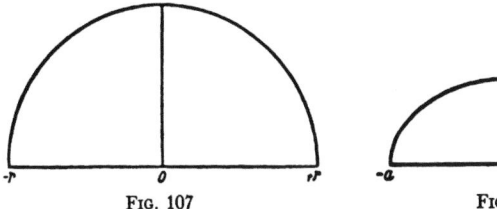

FIG. 107   FIG. 108

$\sin\varphi)\sin\varphi d\varphi$; that is again of the type already discussed. The strength of our quasi-mechanical method is that it disposes with one stroke of a tremendous variety of integrals.

This success naturally arouses the desire to try more general expressions than a^2-x^2 under the radical sign. Let us first take $\int R(x,\sqrt{ax+b})dx$. Placing $\sqrt{ax+b}=u$, $ax+b=u^2$, $x=(1/a)(u^2-b)$, we have $adx=2u\,du$; this integral takes on the form

$$\int R\!\left[\frac{1}{a}(u^2-b),\,u\right]\!\frac{1}{a}\,2u\,du=\int R^*(u)\,du\,,$$

which can be integrated.

Next we consider $\int R(x,\sqrt{Ax^2+Bx+C})dx$. Using again the formula

$$A\,x^2+Bx+C=A\left[\left(x+\frac{B}{2A}\right)^2-\frac{B^2-4AC}{4A^2}\right],$$

and placing $x+(B/2A)=t$, and $(B^2-4AC)/4A^2=\delta^2$, and assuming that $B^2-4AC\geq0$, we obtain

$$\int R(x,\sqrt{a^2-x^2})\,dx=\int R\!\left[t-\frac{B}{2A},\,\sqrt{A(t^2-\delta^2)}\right]dt.$$

For $A<0$ and $B^2-4AC>0$ we have the special case

$$\int R(x,\sqrt{a^2-x^2})\,dx\,,$$

which we already discussed. The case $B^2 - 4AC = 0$ is readily disposed of, for here the square root can be extracted, and we have thus an integral of a rational function. But there remain three further cases:

$$A > 0, \ B^2 - 4AC > 0 \,; \ A > 0, \ B^2 - 4AC < 0 \,; \ A < 0, \cdot B^2 - 4AC < 0 \,.$$

$A = 0$ means that we have the previously discussed case $\sqrt{Bx + C}$, and hence this may be omitted here. Thus, of the four equally possible cases, we have in fact disposed of only one.

All the others can be treated similarly, but we shall not go into the details. We wanted only to give an idea of the painstaking distinctions which have to be made among the various cases and which we have encountered already in the integration of rational functions but which became far more numerous here.

35. LIMITATIONS OF EXPLICIT INTEGRATION

What if an expression of higher than the second degree stands under the radical sign? Or if roots higher than square roots occur? The problem of finding the circumference of the ellipse was one of the first leading to such a situation.

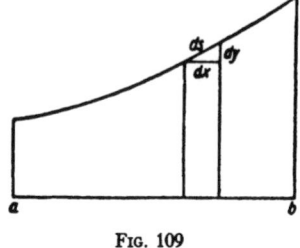

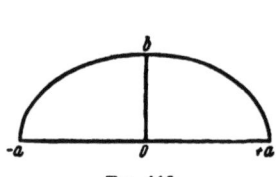

FIG. 109   FIG. 110

How do we express the length of an arc of a curve which is the geometric representation of a function $f(x)$ in an interval ab? Let us proceed, for once, in true Leibnizian fashion: the arc is composed of all the little pieces ds which are almost straight (Fig. 109). According to the Pythagorean theorem

$$(ds)^2 = (dx)^2 + (dy)^2 = \left[1 + \left(\frac{dy}{dx}\right)^2\right](dx)^2.$$

Hence the whole arc is

$$\int_a^b ds = \int_a^b \sqrt{1 + y'^2}\,dx\,.$$

All this can, of course, be derived with greater mathematical fastidiousness, but at the moment we are not interested in that. (This is a fourth item which we defer until later.)

The upper half of the ellipse (Fig. 110) is the geometrical representation of the function $y = (b/a)\sqrt{a^2 - x^2}$. Its circumference, therefore, is $\int_{-a}^{a}\sqrt{1 + y'^2}\,dx$.

From $(x^2/a^2) + (y^2/b^2) = 1$ we obtain $(2x/a^2) + (2y/b^2)y' = 0$; hence $y' = -(x/y)(b^2/a^2)$. The full circumference of the ellipse is, therefore, given by the integral

$$U = 2\int_{-a}^{a}\sqrt{1 + \frac{x^2}{y^2}\frac{b^4}{a^4}}\, dx = 2\int_{-a}^{a}\sqrt{\frac{a^4y^2 + b^4x^2}{a^4y^2}}\, dx = 2\int_{-a}^{a}\frac{\sqrt{a^4y^2 + b^4x^2}}{a^2y}\, dx,$$

or, since $x^2b^2 + a^2y^2 = a^2b^2$, and hence $a^4y^2 = a^4b^2 - a^2b^2x^2$,

$$U = 2\int_{-a}^{a}\frac{\sqrt{a^4b^2 + (b^4 - a^2b^2)x^2}}{ab\sqrt{a^2 - x^2}}\, dx$$

$$= 2\int_{-a}^{a}\frac{a^4 + (b^2 - a^2)x^2}{a\sqrt{a^2 - x^2}\sqrt{a^4 + (b^2 - a^2)x^2}}\, dx$$

$$= 2\int_{-a}^{a}\frac{a^4 + (b^2 - a^2)x^2}{a\sqrt{(a^2 - x^2)[a^4 + (b^2 - a^2]x^2}}\, dx.$$

This indefinite integral has the form

$$\int R[x,\ \sqrt{a^6 - a^4x^2 + (b^2 - a^2)a^2x^2 - (b^2 - a^2)x^4}]dx\ ;$$

that is, under the radical stands a polynomial of degree 4.

For $b = a = r$, for which the ellipse is a circle, the integral reduces to

$$U = \int_{-r}^{r}\sqrt{\frac{x^2 + y^2}{y}}\, dy = 2\,r\int_{-r}^{r}\frac{dx}{\sqrt{1 - x^2}},$$

which is at once found to be $U = 2\pi r$. But for the general case $a \neq b$ the integral resisted all efforts similar to those discussed in the last section. Only in the course of the nineteenth century did Weierstrass and Riemann uncover the deeper reasons why this is not so simple.

The problem to find the circumference of the lemniscate leads to integrals of the same type, as does also the theory of the pendulum, which we are going to treat later on. So far we wish only to push on to the limits of the conquered territory, to show how far out these limits lie, and what a variety of functions we are able to integrate either by special and sometimes tricky devices or, more importantly, by systematically applicable rules.

In the last sections we consciously abandoned the role of the discoverer. We aimed throughout at basic clarity of the concepts, which was not in keeping with seventeenth-century developments. Nevertheless, to the extent that we succeeded in depicting the abundance and variety of the problems and the ease with which they were solved, we shall have given a rather accurate account of the historical situation in this age of the great discoveries.

But we should now, indeed, take a look at the contributions made by the individual discoverers and at the much-debated relations among them. Clarification of these vehement controversies of the past may serve also to clarify the subject matter itself—not as though the mathematical facts could be affected by such an inquiry, for mathematical truths are immutable. But the organization of these truths in our minds, our relation to them, and the use we make of them may indeed be affected by our recognizing their proper ordering.

We recall that Barrow was in possession of most of the rules of differentiation, that he could treat many "inverse tangent problems" (indefinite integrals, we would say), and that in 1667 he discovered and gave an admirable proof for the fundamental theorem—that is, the relation to the definite integral. In his Preface he acknowledges the help given him by his student Isaac Newton. Barrow had originally been a theologian. Later on problems of the calendar and of biblical chronology led him to take an interest in astronomy; and, since this could not be done without mathematics, he began to study Euclid and the other Greek mathematicians and eventually had become the teacher and researcher in mathematics when we encountered him. But then a strange thing happened: After having published the *Lectiones geometricae*, which contained all the discoveries mentioned above, he gave up his teaching position, handing it over to young Newton, while he himself returned to his clerical career, at first in humble circumstances, but gradually rising to a leading position in the Anglican church. Only in his leisure hours did he still occupy himself with Euclid and the other ancients.

In the meantime Newton continued to develop the infinitesimal calculus in various directions. He infused it with the theory of infinite series, in which he had made his first discoveries, and used it in connection with his discovery of the law of gravitation. But in his publications on this law he circumvented the infinitesimal calculus and published nothing on the calculus itself. A manuscript dealing with it he gave to his friend Collins in London. Young Leibniz, student of Huygens, was in London in 1672 and again in 1676, saw the manuscript at Collins', and took notes from it. He himself was already in possession of the differential calculus, the first suggestions of which he had found in Pascal. But it is all in his own conception, bold and unrigorous, in differentials. He too did not publish anything for quite a while. Only when year after year passed without Newton's publishing anything did Leibniz in 1684 begin publishing his results.[41]

Then much later, only after 1700, that unhappy priority dispute broke out which embittered the declining years of both men. It was conducted by their respective disciples with greatest vehemence as a matter of national rivalry, which had a political backdrop. Newton was a member of the House of Lords, held public office, and was, although not in any way active, a Tory. Leibniz, on the other hand, was in the service of the king of Hanover and highly active in politics for this king who pursued his political ambitions in England through the Whig party.

We are merely interested in the question of the substance of the dispute. The basic discoveries could all be found in Isaac Barrow's work published in 1669. But the mathematical public had learned the new calculus in the form published by Leibniz and worked with his differentials. In fact, a "mathematical public" came into being only through these publications. For prior to him, to solve tangent problems or, worse than that, to inverse tangent problems had been an art difficult to master. But now almost anyone could learn these easy and smooth rules, this "calculus"; and when all this began to sail under the flag of "Leibniz," English mathematicians recalled that Barrow and Newton had already been in possession of it. Thus the quarrel started. It does not interest us here for its own sake but only where its true roots were.

They were in the development of mathematics as a whole, especially in the development of the function concept. In our preceding analysis of the origin of the number concept we showed how, apart from Greek mathematics, which was geometrical in form, there developed the computational mathematics of the Babylonians and Indians and how during the Middle Ages—at about the time the Arabs were the guardians of the cultural tradition—both kinds of mathematics began to be amalgamated. When, beginning around 1250, and more completely in the sixteenth century, the West assumed leadership, it took over geometry and computation in a state partly separate, partly fused, partly confused. What ingredients the West threw into the brew at this time we shall not try to discuss; historical research has not yet clarified the matter sufficiently.

What interests us here is, fortunately, a later phase of the development. Viète[42] developed algebra; Descartes,[43] analytic geometry. A large part of Greek geometry had thus become calculational. But Descartes had deliberately refrained from extending his method to the whole of Greek mathematics; he did not touch the infinite processes. It fell to the generation after him to create from this problem the branch of mathematics which we now call "analysis." (This name is a strange and unaccountable transfer to a term which has nothing to do with its original meaning—a counterpart to construction and proof in problems of geometrical constructions.) Now the function concept arises. And it is here that we have to take a specially close look.

The function concept originated in two distinct forms, and this dichotomy is significant for the understanding of its later development—in fact, even for its understanding today. It developed as a geometrical function concept, on the one hand, and as a computational function concept, on the other. We encountered in Galileo and Cavalieri and, in purest form, in Barrow the germs of the geometrical function concept. The function is an abstraction obtained from geometrical and mechanical models, comprising both of them under a conceptual generalization; the function is conceived as a rule which assigns to every x in an interval $a \leq x \leq b$ a number $y = f(x)$. The function concept which, on the other hand, developed out of Viète's theory of equations and Descartes's analytic geometry is the function as a computational expression, a far narrower function concept.

Polynomials, rational functions, radicals, infinite series—as long as the question of their convergence was not given much consideration, their applicability was of course questionable and, in doubtful cases, narrowly construed. This was the computational function concept. While Cavalieri's "indivisibles" were obviously of doubtful validity as long as they were used in connection with the general, geometrical function concept, as "differentials" they became highly viable in the hands of large numbers of mathematicians using the limited, computational function concept. Recall in this connection all the cautious reservations we had to make in Section 26. They are needed only when dealing with the broadest possible concept of a differentiable function. Given a mere computational expression, monstrosities such as we troubled ourselves about can never arise, and we may safely intrust ourselves to the differentials. And this, in fact, was the development which set in in 1640. Gregorius a Santo Vincento, the Jesuit, and Huygens still thought in the ancient Greek manner. Barrow was seized by the current toward the computational function concept and tried to fit it into the Greek mode of thinking. Newton was born only a few years later than Barrow, but he belonged to the new generation of the computational expression. It is strange how much significance there is in the year of birth. In fact, the recognition of the significance of "generations," which we owe to the history of art, is most important also for understanding the history of mathematics. The case before us is a striking instance. What Newton absorbed from the beginning remained foreign to Barrow throughout his life: the turn from the geometrical to the computational function concept—the turn from the confines of the Greek art of proof to the easy flexibility of the indivisibles. On one page of his work Barrow alluded briefly to these matters, but quickly, as though in horror, he dropped them again.

Hence the border line which separates Barrow from his successors lies in the distribution of emphasis. Barrow, it is true, discovered the fundamental theorem. But what did he do with it? In principle, we can do two things with it: We can use known definite integrals like $\int_a^b x^n dx$ to derive from them indefinite integrals like $\int x^n dx$, or we can use known indefinite integrals to obtain definite integrals in accordance with the fundamental theorem. Barrow did more of the former than of the latter. Especially is the proof of the existence of the indefinite integral thus deduced from the existence of the definite one—an existence proof which is still valid in modern mathematics. But it was only his followers who discovered the great ease with which the indefinite integrals of algebraic expressions could be obtained. It is this development which we took as the starting point in our discussion. This had been understood only by the new generation—Newton and Leibniz—and, of the two, Leibniz doubtless did this with greater ease and ingenuity than Newton. The fundamental theorem underwent thereby a change in emphasis which altered its very meaning. Barrow had the theorem, but to him it was not the tool as we use it today. We may speculate that it is somewhat absurd for the priority dispute to have raged between Newton and Leibniz rather than between Barrow and his successors. Barrow himself had left the stage. It was the change of generation which appears to have been the subconscious reason for his resigning his position. In his leisure hours he read only the Greeks—that tells the story. Newton had every reason to argue against Barrow. If Barrow and Newton had been of different nationality, rather than Leibniz and Newton, the quarrel might possibly have broken out between the disciples of Barrow and those of Newton. Objectively, there was far more justification for that.

The verdict of the investigating committee which the Royal Academy had appointed (Newton was one of its members), according to which Leibniz had plagiarized everything, is absurd. The form of Leibniz' calculus, with its easy flexibility, has such perfection that it cannot but have been conceived independently in one great vision. This can also be established from his notes and letters. Only problematic is his relation to the fundamental theorem. If the dispute turned later—and in fact still does so today—into entirely wrong directions, it was because people failed to present the role of the fundamental theorem as clearly as we have done here and as was first suggested by Felix Klein in a lecture course given in Göttingen. From the notes in our possession we cannot tell with certainty whether Leibniz *recognized* the fundamental theorem and translated it into his own notation or whether he *discovered* it. We have, in Hanover, Leibniz' library as used by him and the papers he left. Among his books are the *Lectiones geometricae* by Barrow. He had the habit when reading to underscore important passages with ruler and pencil. There are many such underscorings in the first *Lectiones* of Barrow. But then they stop, long before the chapters which interest us here, although there are marginal notes in these. Researchers have tried to prove that these date from later years. All these efforts seem pointless as a defense of Leibniz. He stands in no need of them, nor do they do anything for him. Barrow's work was in print and known to all contemporaries. Leibniz entertained uncommonly extensive and eager communication with him, personally and by way of correspondence. Still it is not clear whether he knew Barrow's theorem. The priority is, of course, Barrow's. The shift in emphasis in this theorem, of which we spoke above, was brought about by both Leibniz and Newton. In the manner, however, in which this was achieved, Leibniz was so much clearer and more perfect that his independence in this matter cannot be doubted at all. For that no proof is necessary.

Thus the debate over this dispute, down into our days, has suffered from the fact that, owing to a lack of mathematical understanding, the question was posed in the wrong way. And that is why it may be hoped that the analysis of this dispute here has contributed to our own mathematical understanding.

The computational function concept remained dominant throughout the eighteenth century and celebrated triumphs in the great discoveries of Euler, the Bernoullis, Lagrange, and others. When around 1820 Fourier, prompted by the needs of physics, found it necessary to resurrect the geometrical function concept—the concept of the "arbitrary function"—the difference between the two was again brightly illuminated. This time it was the "old generation"—

Cauchy, Poisson, Laplace, etc.—who stood up for the computational function concept and who assailed Fourier series, which represent functions with a break, functions which follow different laws to the left and right of such breaks. But again the new generation was victorious and, without being aware of it, picked up the thread where Barrow had left off. The whole nineteenth century was needed to develop both fully—computational expression and function—in conjunction with and independently from each other, each according to its peculiarity. Today's researchers have them both at their disposal. They use them separately or in mutual interpenetration. For the student, however, it is difficult to keep them apart; the textbooks he studies do not give him enough help, because they tend to blur rather than to sharpen the difference.

4

APPLICATIONS TO PROBLEMS OF MOTION

36. VELOCITY AND ACCELERATION

So far we have become acquainted with only one kind of motion—that of the freely falling body, with a brief glance also at the fall on an inclined plane. We have thereby become acquainted with motion in a straight line only, and from it we have derived the idea of the mechanical interpretation of a function $s = f(t)$. Its derivative $ds/dt = f'(t)$ gives the velocity. We also considered the second derivative $d^2s/dt^2 = f''(t)$, which was a constant g in the case of the freely falling body. It is constant also in the motion on the inclined plane, as proved

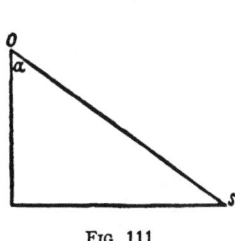

Fig. 111

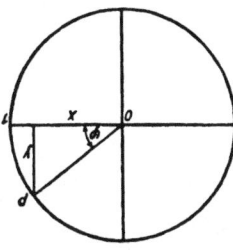

Fig. 112

by Galileo's experiments, but the constant was smaller the larger the angle between the inclined plane and the vertical. More precisely, Galileo found that $d^2s/dt^2 = g \cos \alpha$ (Fig. 111).

But how can we mathematically describe a motion which does not take place along a straight line? Let us, for example, take one of the simplest motions that can be imagined: a point moves with constant speed in a circle of radius 1 (Fig. 112). Its position in the plane of the circle at any moment can be described by two coordinates, say, within a coordinate system which has the center of the circle as its origin and relative to which the equation of the circle is $x^2 + y^2 = 1$. From instant to instant the position of the point and hence its coordinates x and y vary; both are functions of the time t, $x(t)$ and $y(t)$. In fact, in our case, in which we assumed the angle to vary uniformly in time, $\varphi = ct$, we have simply

$$x = \cos(ct), \qquad y = \sin(ct).$$

It is clear that, conversely, every pair of functions $x(t)$ and $y(t)$ can be interpreted as a definite motion of a point $P(x, y)$ in the plane. By the same token, any motion in space is described by a triple of functions, $x(t)$, $y(t)$, and $z(t)$.

How, then, do we express the "velocity" of a motion given by the equations $x = x(t)$, $y = y(t)$? At a time t_1 the point P is at $x_1 = x(t_1)$, $y_1 = y(t_1)$. From the instant t it has therefore moved along the curve by $\sqrt{(x_1 - x)^2 + (y_1 - y)^2}$ (Fig. 113). The closer to t we take t_1, the closer does the expression

$$\frac{PP_1}{t_1 - t} = \frac{\sqrt{(x_1 - x)^2 + (y_1 - y)^2}}{t_1 - t} = \sqrt{\left(\frac{x_1 - x}{t_1 - t}\right)^2 + \left(\frac{y_1 - y}{t_1 - t}\right)^2}$$

approach $\sqrt{[x'(t)]^2 + [y'(t)]^2}$, or, written more briefly, $\sqrt{x'^2 + y'^2}$. This, then, is the "absolute value" of the velocity ds/dt, or, as it is called, its speed. Its "di-

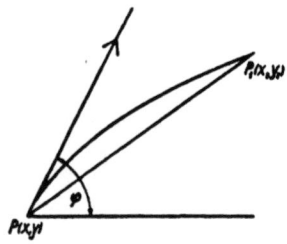

FIG. 113

rection" is given by the angle φ which the tangent line drawn from P in the direction of the motion forms with the positive x-axis.

For the determination of φ we have $\tan \varphi = dy/dx$. But this is really insufficient, because $\tan \varphi$ determines φ uniquely only between zero and π. Hence we had better use $\sin \varphi$ and $\cos \varphi$. These result, respectively, as the limiting values of the expressions

$$\frac{x_1 - x}{\sqrt{(x_1 - x)^2 + (y_1 - y)^2}} = \frac{(x_1 - x)/(t_1 - t)}{\sqrt{\left(\frac{x_1 - x}{t_1 - t}\right)^2 + \left(\frac{y_1 - y}{t_1 - t}\right)^2}}$$

and

$$\frac{y_1 - y}{\sqrt{(x_1 - x)^2 + (y_1 - y)^2}} = \frac{(y_1 - y)/(t_1 - t)}{\sqrt{\left(\frac{x_1 - x}{t_1 - t}\right)^2 + \left(\frac{y_1 - y}{t_1 - t}\right)^2}}$$

$$\cos \varphi = \frac{x'}{\sqrt{x'^2 + y'^2}}, \qquad \sin \varphi = \frac{y'}{\sqrt{x'^2 + y'^2}}.$$

For the uniform circular motion ($c > 0$) we have $x' = -c \sin ct$, $y' = c \cos ct$; $x'^2 + y'^2 = c^2$, $\sqrt{x'^2 + y'^2} = c$. That is, the factor of proportionality c gives the absolute value of the velocity. Its direction is given through

$$\cos \varphi = -\frac{c \sin ct}{c} = \sin(-ct) = \cos\left(\frac{\pi}{2} + ct\right),$$

$$\sin \varphi = \frac{c \cos ct}{c} = \cos(-ct) = \sin\left(\frac{\pi}{2} + ct\right);$$

that is, through $\varphi = (\pi/2) + ct$, as it should be.

Likewise, as we described the velocity of the motion $x(t)$, $y(t)$ by giving its absolute value $\sqrt{x'^2 + y'^2}$ and its direction by

$$\cos \varphi = \frac{x'}{\sqrt{x'^2 + y'^2}}, \qquad \sin \varphi = \frac{y'}{\sqrt{x'^2 + y'^2}},$$

we shall describe the "velocity of the velocity," that is, the acceleration, by giving its absolute value $J = \sqrt{x''(t)^2 + y''(t)^2}$ and its direction by

$$\cos \psi = \frac{x''}{\sqrt{x''^2 + y''^2}}, \qquad \sin \psi = \frac{y''}{\sqrt{x''^2 + y''^2}}.$$

These two magnitudes, to repeat, are obtained from $x'(t)$ and $y'(t)$ by the same principle by which the absolute value and the direction of the velocity were obtained from $x(t)$, $y(t)$. It should be noted that the derivative $(d/dt)(ds/dt) = (d^2s/dt^2)$ of which we might think in this connection,

$$\frac{d^2 s}{dt^2} = \frac{d}{dt}(\sqrt{x'^2 + y'^2}) = \frac{d}{dt}(x'^2 + y'^2)^{1/2}$$

$$= \tfrac{1}{2}(x'^2 + y'^2)^{-1/2}[2x'x'' + 2y'y''] = \frac{x'x'' + y'y''}{\sqrt{x'^2 + x'^2}},$$

is emphatically not the same as J.

For instance, in the case of uniform circular motion, we have

$$x'' = -c^2 \cos ct, \qquad y'' = -c^2 \sin ct,$$

$$J = \sqrt{x''^2 + y''^2} = c^2,$$

$$\cos \psi = \frac{x''}{\sqrt{x''^2 + y''^2}} = -\cos ct = \cos(ct + \pi),$$

$$\sin \psi = \frac{y''}{\sqrt{x''^2 + y''^2}} = -\sin ct = \sin(ct + \pi);$$

that is, $\psi = ct + \pi$. Although $d^2s/dt^2 = 0$, $J = c^2$, and the direction of the acceleration is that of the radius, not of the tangent. A constant acceleration is thus imparted which lies normal to the path of the moving point, pointing inward, perpendicular to the circle. When later on we shall discuss *what* it is that imparts this acceleration, we shall see its "cause" in the "centripetal force," given, for example, by the tension of the string which holds the revolving mass point in its circular path.

These findings can be generalized. If a point moves in any path with constant speed, the acceleration is always directed perpendicular (normal) to the path. For, if $d^2s/dt^2 = 0$, we have, as found above, $x'x'' + y'y'' = 0$, or

$$\frac{y''}{x''} = -\frac{1}{x'/y'}, \quad \text{or } \tan \psi = \frac{\sin \psi}{\cos \psi} = \frac{y''}{x''} = -\frac{1}{\tan \varphi} = -\cot \varphi.$$

That is, φ and ψ are directions perpendicular to each other. Now let us place $(x''/y') = -(y''/x') = u > 0$ (otherwise we would say $y''/x' = v > 0$ and proceed analogously). Then

$$J = \sqrt{u^2 y'^2 + u^2 x'^2} = u \sqrt{x'^2 + y'^2} = u \frac{ds}{dt},$$

while ds/dt, the speed, is constant by hypothesis. Thus J is constant if and only if u is constant. Now $x'' = uy'$ and $y'' = -ux'$. Hence $x''y' - y''x' =$

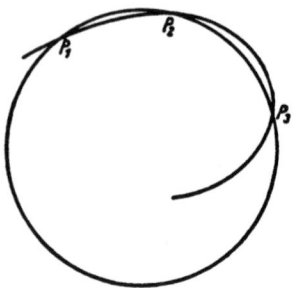

Fig. 114

$u(x'^2 + y'^2) = u(ds/dt)^2 = J(ds/dt)$. In a subsequent section on curvature we shall develop the following ideas: Let P_1, P_2, and P_3 be three points on a curve (Fig. 114). Through these, as through any three points not on a straight line, we can pass one and only one circle. Now, if we hold P_1 fixed, and let P_2 and P_3 move along the curve closer and closer to P_1, the circle determined by them varies too but will approach a limiting position. This limit circle is called the "circle of curvature" to the curve at P_1; its radius r is called the "radius of curvature"; and the reciprocal $1/r = k$ is called "the curvature." In the section on curvature we shall show that

$$k = \frac{x'y'' - y'x''}{(x'^2 + y'^2)^{3/2}}, \qquad k = -\frac{x''y' - x'y''}{\sqrt{x'^2 + y'^2}} \frac{1}{x'^2 + y'^2} = -\frac{J}{(ds/dt)^2}.$$

Now, if ds/dt is constant, as had been assumed, J is proportional to the curvature. Hence J will be constant if and only if the curvature is constant, that is, if the curve is a circle. Otherwise, the sharper the curve, the greater is J.

What is the "cause" of the acceleration? Let us think, for example, of a train which travels with constant speed along a curve. In that case it is the "strain of the rails" which keeps the train from running straight ahead in the direction of the tangent. This "strain" is proportional to the curvature of the rails.

In the case $k = 0$, that is, if the curve is a straight line, $J = 0$. In uniform motion along a straight line the acceleration is zero. The converse of this statement is also true: If in a plane motion the acceleration is zero, $x''^2 + y''^2 = 0$. For the sum of two squares, neither of which can be negative, to be zero means that both are zero, that is, $x'' = 0, y'' = 0$. Hence $x' = a, y' = b$, where a and b are constants, and therefore $\sqrt{x'^2 + y'^2} = \sqrt{a^2 + b^2}$, which is constant. It means further that $x = at + a, y = vt + \beta$, and therefore $vx - ay = ab - \beta a;$ that is, x and y satisfy the equation of a straight line—the motion is rectilinear.

For Galileo this was the point of departure: in the absence of acceleration, the point moves with uniform velocity in a straight line. He called this the "law of inertia." For us it is a mathematical theorem which we proved. Only in speaking of the "cause" of the acceleration as a "force" which is acting upon the body, implying that a body on which no "force" acts moves uniformly in a straight line, we leave the domain of pure mathematical assertion and are obliged, first of all, to explain the concept of "force."

"Acceleration" is a complex, yet subtly mathematical concept, free from any ambiguity. With the word "force," we associate an immediate intuitive knowledge: we often generate motions and experience the amount of "force" which we have to exert with our own body. "Force" is an idea immediately understood but lacking definition and precision. How are "force" and "acceleration" related to each other? Is force simply the same as acceleration? We see at once that this is not so. Let us think again of the train running on the curve, and let us imagine another train, but a much heavier one, traveling behind it through the same curve, with the same speed. "Mathematically," the motion is exactly the same; hence the acceleration is the same. But we know well enough that the heavier train going through the curve exerts a greater strain or stress on the rails. Hence force and acceleration are not simply identical. However, the clarification of this question came from another side where it could be made more easily. We shall therefore postpone it temporarily and stay deliberately with the purely mathematical concept of acceleration.

So far we considered a given motion and inquired into its acceleration, assuming that every motion has a definite acceleration with a definite absolute value and direction. But if the acceleration is always zero, what kind of motion does that imply? Obviously, such a condition does not yet completely determine the motion; it only means that the motion is rectilinear and uniform. Where the moving point was at time $t = 0$ and what velocity it had at that time are in no way determined. If, however, we indicate in addition the position of the point at $t = 0$, that is, $x(0)$ and $y(0)$, and also the direction and absolute value of its velocity at $t = 0$, that is, $x'(0)$ and $y'(0)$, then indeed the motion is fully determined.

Another problem of this kind, which Newton also found solved in Galileo,[44] is the case of the freely falling body, the acceleration of which is constant and directed downward. Let there be a vertical plane with the y-axis pointing down and a point P moving with an acceleration of constant absolute value g directed downward. This means that $x'' = 0$, and $J = \sqrt{x''^2 + y''^2} = \sqrt{y''^2} = y'' = g$. It follows that

$$y' = gt + a , \quad y = \tfrac{1}{2}gt^2 + at + b ,$$

$$x' = A , \quad x = At + B .$$

In case $A = 0$, which means $x' = 0$, the initial velocity is directed vertically downward or upward; hence $x = B$, that is, constant. The motion, therefore, takes place in a vertical line and is determined by the initial position $y(0) = b$ and the initial velocity $y'(0) = a$ (the vertically projected body).

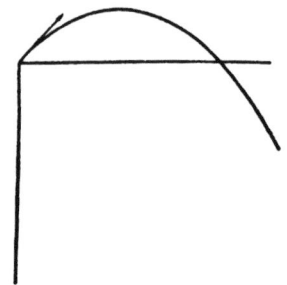

FIG. 115

In case $A \neq 0$ (motion of a projectile) we have

$$At = x - B , \quad t = \frac{1}{A} x - \frac{B}{A} ,$$

$$y = \tfrac{1}{2} g \left(\frac{1}{A} x - \frac{B}{A} \right)^2 + a \left(\frac{1}{A} x - \frac{B}{A} \right) + b$$

$$= \tfrac{1}{2} g \frac{1}{A^2} x^2 + \left(\frac{a}{A} - \tfrac{1}{2} g \frac{2B}{A^2} \right) x + \left(\tfrac{1}{2} g \frac{B^2}{A^2} - a \frac{B}{A} + b \right)$$

$$= ux^2 + vx + w .$$

This means that the projectile travels through a parabolic path whose maximum height and range can be readily computed (Fig. 115).

37. THE PENDULUM

Christian Huygens, great Dutch mathematician and physicist of the seventeenth century and teacher of Leibniz, who anticipated many of the ideas of infinitesimal calculus but disliked the *Kalkül*, was also the inventor of the pendulum clock, which assumed such great importance for practical and scientific purposes.[45] To Galileo it was still very difficult to measure the "equal times" in which his lead balls rolling down the inclined planes passed the markings at distances 1, 4, 9, . . . , length units. He invented for this purpose various water clocks which functioned on the principle of the sandglass* still in use today for boiling eggs; the amount of water gone from the vessel was a measure of the lapse of time.

The motion made by a suspended mass point swinging back and forth around its equilibrium position cannot be altogether understood by mere observation. Huygens tried to analyze it on the basis of the following consideration: Galileo, in observing the fall on the inclined plane with the angle a, had found that $d^2s/dt^2 = g \cos a$ and is thus independent of t, while g had the same value for

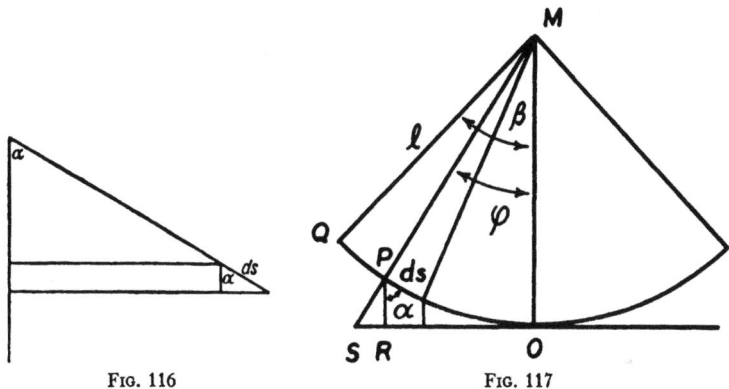

FIG. 116   FIG. 117

all a and hence the same value as for the freely falling body ($a = 0$). Its value is approximately 10, if we measure time in seconds and lengths in meters (Fig. 116).

Let us now imagine the point P of the pendulum, constrained by the string on which it is suspended to stay on the circular orbit $x^2 + y^2 = l^2$, traversing in the short time interval dt the almost straight little distance ds along the circle and having the acceleration $g \cos a$, as if it were rolling down an inclined plane with the angle a (Fig. 117). (We cannot say how closely this hypothesis agrees with nature. But we shall derive mathematical consequences from it which can be tested to a high degree of accuracy. The mathematics to be applied will be largely taken from our previously acquired knowledge.)

Angle a, which ds forms with the vertical, is the complement of angle SPR. The latter, in turn, is equal to φ (as alternate angle on parallel lines); hence $a = (\pi/2) - \varphi$ and $\sin \varphi = \sin [(\pi/2) - a] = \cos a$. Thus $d^2s/dt^2 = g \sin \varphi$. On the other hand, the circular arc s from Q to P (where Q is the point from which

*a small hourglass

we release the pendulum at angle β without giving it any velocity) is $s = l(\beta - \varphi)$, all angles being measured in radians. Hence

$$\frac{ds}{dt} = -l\frac{d\varphi}{dt}, \qquad \frac{d^2s}{dt^2} = -l\frac{d^2\varphi}{dt^2},$$

$$l\frac{d^2\varphi}{dt^2} = -g \sin \varphi. \tag{37.1}$$

Thus our hypothesis concerning the nature of motion has led to the statement that $\varphi(t)$ satisfies equation (37.1). Such an equation is called a "differential equation," holding between an unknown function—in this case $\varphi(t)$—and its derivatives.

So far, we have met with only one species of differential equations—those of the form $dy/dx = f(x)$. Their solutions are simply of the form $y = {}^x\!\int_a f(t)dt$. Our present differential equation is more difficult. But by a clever device we shall reduce it to a manageable integration. We multiply both sides of (37.1) by φ':

$$l\varphi'\varphi'' = -g\varphi' \sin \varphi.$$

Now $\varphi'\varphi'' = (d/dt)(\frac{1}{2}\varphi'^2)$, and $(d/dt)(-\cos \varphi) = \varphi' \sin \varphi$. Thus the two functions $l\frac{1}{2}\varphi'^2$ and $g \cos \varphi$ obviously have the same derivative and hence, according to the fundamental theorem, differ only by a constant: $l\frac{1}{2}\varphi'^2 = g \cos \varphi + c$.

How do we determine this constant? We must know the exact $\varphi'(t)$ for at least one value of t. That we do know for $t = 0$, at which moment (taking it as the beginning of the motion) we released the pendulum with the amplitude $\varphi = \beta$ but without imparting to it any initial velocity. This means $\varphi'(0) = 0$; and hence $\frac{1}{2}l\cdot0 = g \cos \beta + c$, $c = -g \cos \beta$. Substituting this into our previous equations, we obtain

$$\frac{1}{2}l\varphi'^2 = g(\cos \varphi - \cos \beta),$$

$$\varphi'^2 = \frac{2g}{l}(\cos \varphi - \cos \beta),$$

$$\frac{d\varphi}{dt} = \sqrt{\frac{2g}{l}} \sqrt{\cos \varphi - \cos \beta}. \tag{37.2}$$

We have thus succeeded in removing the second derivative, but our new differential equation is still far from the form $dy/dx = f(x)$; rather it has the form $dy/dx = f(y)$. But that can easily be remedied, for $dx/dy = 1/f(y)$, an equation of the desired form. Hence $x = \int(dy)/[f(y)]$ is the solution, and in this case we may obtain

$$\frac{dt}{d\varphi} = \sqrt{\frac{l}{2g}} \frac{1}{\sqrt{\cos \varphi - \cos \beta}},$$

$$t = \sqrt{\frac{l}{2g}} \int \frac{d\varphi}{\sqrt{\cos \varphi - \cos \beta}},$$

which is a trigonometric integral. Only the integrand is not a rational function $R(\cos \varphi, \sin \varphi)$ because of the square root. To transform it, we place

$$u = \frac{\sin \frac{1}{2}\varphi}{\sin \frac{1}{2}\beta},$$

$$\frac{du}{d\varphi} = \frac{\cos \frac{1}{2}\varphi}{2 \sin \frac{1}{2}\beta},$$

or

$$\frac{d\varphi}{2 \sin \frac{1}{2}\beta} = \frac{2\,du}{\cos \frac{1}{2}\varphi}.$$

However, since $\cos 2x = 1 - 2 \sin^2 x$,

$$\cos \varphi = 1 - 2 \sin^2 \frac{\varphi}{2},$$

$$\cos \beta = 1 - 2 \sin^2 \frac{\beta}{2},$$

and

$$\cos \varphi - \cos \beta = 2 \left(\sin^2 \frac{\beta}{2} - \sin^2 \frac{\varphi}{2} \right).$$

Therefore,

$$t = \sqrt{\frac{l}{2g}} \int \frac{d\varphi}{\sqrt{2 \left[\sin^2(\beta/2) - \sin^2(\varphi/2)\right]}} = \sqrt{\frac{l}{4g}} \int \frac{d\varphi}{\sqrt{\sin^2(\beta/2) - \sin^2(\varphi/2)}}$$

$$= \frac{1}{2}\sqrt{\frac{l}{g}} \int \frac{1}{\sqrt{1 - \frac{\sin^2(\varphi/2)}{\sin^2(\beta/2)}}} \frac{d\varphi}{\sin(\beta/2)} = \frac{1}{2}\sqrt{\frac{l}{g}} \int \frac{1}{\sqrt{1 - u^2}} \frac{2\,du}{\cos(\varphi/2)}$$

$$= \sqrt{\frac{l}{g}} \int \frac{du}{\sqrt{(1 - u^2)(1 - u^2 \sin^2[\beta/2])}}, \qquad (37.3)$$

since

$$\cos \frac{\varphi}{2} = \sqrt{1 - \sin^2 \frac{\varphi}{2}} = \sqrt{1 - u^2 \sin^2 \frac{\beta}{2}}.$$

(That is the same type of integral which we met before when attempting to find the circumference of the ellipse; we call these "elliptic integrals.")

Postponing (37.3), we return to (37.2), which gives the velocity $d\varphi/dt$ as a function of φ. From it we can deduce several facts as regards the motion (Fig. 118). The velocity $\varphi'(t)$ of the pendulum which begins at $t = 0$ with $\varphi'(0) = 0$ increases, while φ decreases from β to 0. At $\varphi = 0$, $\cos \varphi$ attains its maximum value 1, and hence $\cos \varphi - \cos \beta$ attains its maximum $1 - \cos \beta$; still further, $d\varphi/dt$ attains its maximum

$$\sqrt{\frac{2g}{l}} \sqrt{1 - \cos \beta}.$$

After that, $d\varphi/dt$ decreases, assuming the same value for $-\varphi$ as for φ. At $\varphi = -\beta$ it becomes zero again and then swings back and forth ad infinitum. (All this follows purely mathematically from our hypothesis and may be regarded as a first verification of it.)

We can now compute the time which the pendulum takes to return, for the first time, to its initial position, that is, its "period of oscillation" T. No matter what the indefinite integral may be in (37.3), that is, independent of how we determine the so-far-undetermined constant, according to the fundamental theorem we shall have

$$T = 2 \sqrt{\frac{l}{g}} \int_{-1}^{1} \frac{du}{\sqrt{(1 - u^2)(1 - u^2 \sin^2 \beta/2)}}.$$

This integral, it is true, is still beyond our powers; we can, however, handle a justifiable approximation. In pendulum clocks, angle β is rather small, and

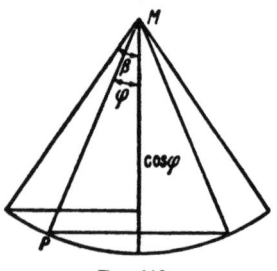

Fig. 118

hence $\sin^2(\beta/2)$ is very small indeed. To see how small it is, we give the following tabulation taken from a cosine table, since $\sin^2(\beta/2) = \frac{1}{2}(1 - \cos \beta)$:

$\beta = 12°$	$\sin^2\frac{\beta}{2} = 0.0109$	$\beta = 6°$	$\sin^2\frac{\beta}{2} = 0.0028$
11°	0.0092	5°	0.0019
10°	0.0076	4°	0.0012
9°	0.0062	3°	0.0007
8°	0.0049	2°	0.0003
7°	0.0038	1°	0.0001

Already at an amplitude 10°, the value of $\sin^2(\beta/2)$ is less than 0.01, and hence

$$1 - \tfrac{1}{100}u^2 \leq 1 - u^2 \sin^2 \frac{\beta}{2} \leq 1;$$

thus T is very close to

$$2\sqrt{\frac{l}{g}} \int_{-1}^{1} \frac{du}{\sqrt{1 - u^2}} = 2\sqrt{\frac{l}{g}} [\text{arc sin } u]_{-1}^{1} = 2\pi\sqrt{\frac{l}{g}}, \qquad (37.4)$$

$$T = 2\pi\sqrt{\frac{l}{g}}.$$

This value is independent of β. For $\beta < 10°$, the value of T will not noticeably change with β but will be practically constant, approaching the limiting value $T = 2\pi\sqrt{l/g}$.

Huygens had considerable scruples about this approximation. Later measurements showed, however, that the neglect of the pendulum friction and the air resistance are far greater than that of $\sin^2 (\beta/2)$, so that it was not worthwhile to spend more effort on the pendulum problem. However, Huygens' further reasoning on this problem led to some of his greatest mathematical achievements; hence a few words about it.

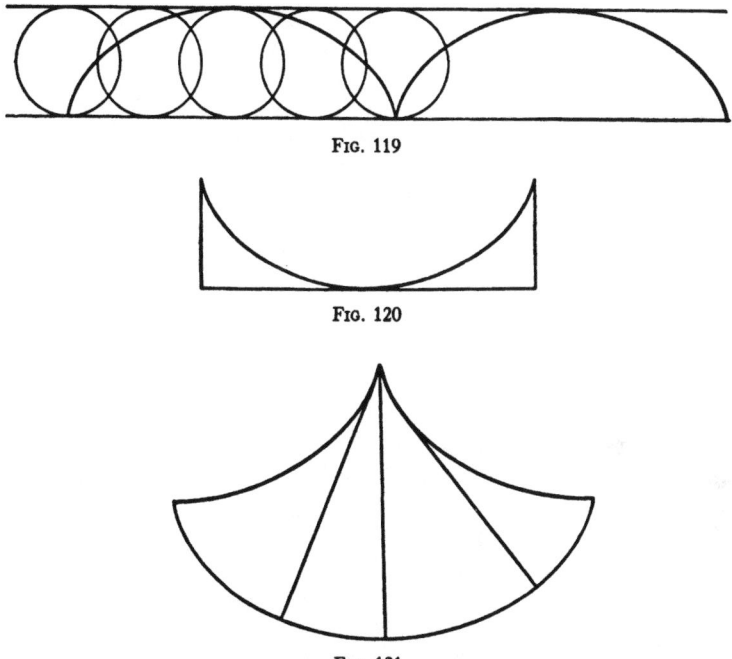

Fig. 119

Fig. 120

Fig. 121

Huygens was interested in a pendulum whose period of oscillation should be strictly independent of its amplitude; and he found it. He asked himself: In what kind of curve other than a circular arc should the pendulum be made to oscillate? He showed that there was only one such curve, the so-called cycloid, which is described by a point, say, a bit of paper, on the circumference of a rolling wheel (Fig. 119). Huygens placed an arc of this curve upside down, and on it he let a ball run back and forth (Fig. 120). Or, rather, he constructed a pendulum whose bob was made to describe this curve, because the suspension string was clamped between two suitable jaws against which it pressed closer the higher the pendulum rose (Fig. 121). The mathematical derivation of its period of oscillation—now strictly constant—is very similar to that of the circular pendulum but only easier. Huygens' great achievement was the proof that the cycloid is the only curve which has this property. His reasoning in that proof goes far beyond differential and integral calculus, reaching into new realms of thought. To this day it stands as a unique achievement.

Let us return to the ordinary pendulum. In the seventeenth century, the pendulum was one of the finest instruments not only for the measuring of time but also for the determination of the general acceleration g, which was but very crudely measurable by experiments on the inclined plane. Next to the telescope, in the invention of which Galileo participated, the pendulum clock has done most to revolutionize astronomy. (Of that, however, we do not want to speak, because the astronomical discoveries to be discussed below belong to a time in which these instruments were not yet used. But the importance of the pendulum for the determination of g does concern us already at this place.)

In 1672 the French astronomer Richer, who had been sent to Cayenne (French Guiana) on a scientific mission, noticed that his pendulum clock, which had kept very accurate time in Paris, was losing more than 2 minutes per day. Newton, who of course was acquainted with Huygens' formula (37.4), inferred from this observation that the acceleration g was smaller in Cayenne than in Paris. We shall see how Newton used this fact.

38. COORDINATE TRANSFORMATIONS

The definition of the magnitude and direction of acceleration gives rise to a question which we should not pass by. We based the definition on a rectangular coordinate system relative to which the motion was described by the pair of functions $x(t)$ and $y(t)$, from which we obtained the expression $\sqrt{\ddot{x}^2 + \ddot{y}^2}$, etc. (From now on we shall denote the time derivatives of the functions by a dot above the function symbol.) But what would happen if we were to use a different rectangular coordinate system (u, v) having a different origin and direction? Will $\sqrt{\ddot{u}^2 + \ddot{v}^2}$ have the same value as $\sqrt{\ddot{x}^2 + \ddot{y}^2}$, and what will be the direction determined by

$$\frac{\ddot{u}}{\ddot{u}^2 + \ddot{v}^2} \quad \text{and} \quad \frac{\ddot{v}}{\ddot{u}^2 + \ddot{v}^2} ?$$

That is not immediately obvious. Yet it would apparently be ruinous for the concept of acceleration if it depended on the coordinate system, for we must demand that the properties of motion pertain to the motion itself and not to an extraneous mathematical device like a coordinate system.

It is not difficult to show the independence from the coordinate system (Fig. 122). Let (ξ, η) be a coordinate system parallel to the (x, y) system but having a different origin,

$$\xi = a + x, \quad \eta = b + y.$$

Next, let (u, v) be a system having the same origin as the (ξ, η) system but rotated relative to it by an angle ϑ. Then

$$u = \xi \cos \vartheta + \eta \sin \vartheta$$
$$= x \cos \vartheta + y \sin \vartheta + a ,$$
$$v = -\xi \sin \vartheta + \eta \cos \vartheta$$
$$= -x \sin \vartheta + y \cos \vartheta + \beta .$$

This gives us the pair of functions $u(t)$ and $v(t)$ which describe the motion relative to the (u, v) system; ϑ, a, and β are all constants independent of t, de-

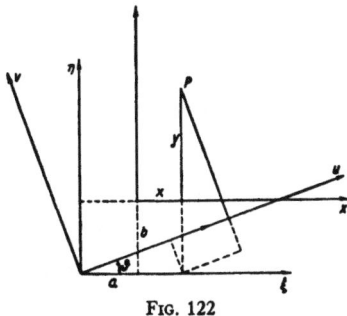

Fig. 122

termined by the position of the new coordinate system relative to the old one. From these we find next

$$\ddot{u} = \ddot{x} \cos \vartheta + \ddot{y} \sin \vartheta ,$$
$$\ddot{v} = -\ddot{x} \sin \vartheta + \ddot{y} \cos \vartheta ;$$

and hence

$$\ddot{u}^2 + \ddot{v}^2 = (\ddot{x}^2 \cos^2 \vartheta + \ddot{y}^2 \sin \vartheta + 2\ddot{x}\ddot{y} \cos \vartheta \sin \vartheta)$$
$$+ (\ddot{x}^2 \sin^2 \vartheta + \ddot{y}^2 \cos^2 \vartheta - 2\ddot{x}\ddot{y} \cos \vartheta \sin \vartheta) = \ddot{x}^2 + \ddot{y}^2 .$$

This shows that the magnitude of acceleration has indeed remained unchanged under this coordinate transformation. What about the direction ψ? We have

$$\frac{\ddot{u}}{\sqrt{\ddot{u}^2 + \ddot{v}^2}} = \frac{\ddot{x} \cos \vartheta + \ddot{y} \sin \vartheta}{\sqrt{\ddot{x}^2 + \ddot{y}^2}} = \cos \psi \cos \vartheta + \sin \psi \sin \vartheta = \cos (\psi - \vartheta)$$

$$\frac{v}{\sqrt{\ddot{u}^2 + \ddot{v}^2}} = \frac{-\ddot{x} \sin \vartheta + \ddot{y} \cos \vartheta}{\sqrt{\ddot{x}^2 + \ddot{y}^2}} = -\cos \psi \sin \vartheta + \sin \psi \cos \vartheta = \sin (\psi - \vartheta) .$$

That is, the new direction of the acceleration is $\psi - \vartheta$ instead of ψ. But this is just as it should be, for if the "direction of acceleration" has remained the

same, it must form an angle diminished by ϑ with the coordinate axis which
was rotated by ϑ.

Let us perform another coordinate transformation which we shall have to
use immediately.

If (x, y) are the rectangular coordinates of a point P, then (r, φ) (see Fig. 123)
are called the "polar coordinates" of the point. To every point P in the plane
such a pair of numbers may be assigned; and, inversely, for $r > 0$ and any
given angle φ, there is one and only one point P in the plane having these as
its polar coordinates, namely, the intersection of the circle drawn with the
radius r around 0 with the half-line from 0 forming the given angle with the
positive direction of the x-axis (Fig. 124).

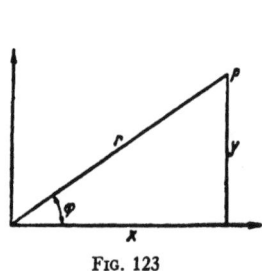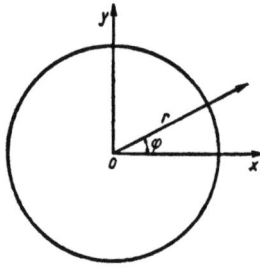

Fɪɢ. 123   Fɪɢ. 124

The relation between the (r, φ) and (x, y) coordinates is at once seen to be

$$x = r \cos \varphi , \qquad y = r \sin \varphi ,$$

$$r = \sqrt{x^2 + y^2}, \qquad \varphi = \text{arc tan } \frac{y}{x}.$$

If now x and y are functions of t describing a motion in the plane, we obtain:

$$\dot{x} = \dot{r} \cos \varphi - r\dot{\varphi} \sin \varphi , \qquad \dot{y} = \dot{r} \sin \varphi + r\dot{\varphi} \cos \varphi ,$$

$$\ddot{x} = \ddot{r} \cos \varphi - \dot{r}\dot{\varphi} \sin \varphi - \dot{r}\dot{\varphi} \sin \varphi - r\ddot{\varphi} \sin \varphi - r\dot{\varphi}^2 \cos \varphi ,$$

$$\ddot{y} = \ddot{r} \sin \varphi + \dot{r}\dot{\varphi} \cos \varphi + \dot{r}\dot{\varphi} \cos \varphi + r\ddot{\varphi} \cos \varphi - r\dot{\varphi}^2 \sin \varphi ,$$

or

$$\ddot{x} = (\ddot{r} - r\dot{\varphi}^2) \cos \varphi - (2\dot{r}\dot{\varphi} + r\ddot{\varphi}) \sin \varphi ,$$

$$\ddot{y} = (\ddot{r} - r\dot{\varphi}^2) \sin \varphi + (2\dot{r}\dot{\varphi} + r\ddot{\varphi}) \cos \varphi ;$$

and hence

$$\ddot{x}^2 + \ddot{y}^2 = (\ddot{r} - r\dot{\varphi}^2)^2 \cos^2 \varphi + (2\dot{r}\dot{\varphi} + r\ddot{\varphi})^2 \sin^2 \varphi$$

$$- 2(\ddot{r} - r\dot{\varphi}^2)(2\dot{r}\dot{\varphi} + r\ddot{\varphi}) \cos \varphi \sin \varphi + (\ddot{r} - r\dot{\varphi}^2)^2 \sin^2 \varphi + (2\dot{r}\dot{\varphi} + r\ddot{\varphi})^2 \cos^2 \varphi$$

$$+ 2(\ddot{r} - r\dot{\varphi}^2)(2\dot{r}\dot{\varphi} + r\ddot{\varphi}) \cos \varphi \sin \varphi ,$$

$$\ddot{x}^2 + \ddot{y}^2 = (\ddot{r} - r\dot{\varphi}^2)^2 + (2\dot{r}\dot{\varphi} + r\ddot{\varphi})^2 , \tag{38.1}$$

$$\left.\begin{array}{l}
\cos\psi = \dfrac{\ddot{x}}{\sqrt{\ddot{x}^2 + \ddot{y}^2}} = \dfrac{(\ddot{r} - r\dot{\varphi}^2)\cos\varphi - (2\dot{r}\dot{\varphi} + r\ddot{\varphi})\sin\varphi}{\sqrt{(\ddot{r} - r\dot{\varphi}^2)^2 + (2\dot{r}\dot{\varphi} + r\ddot{\varphi})^2}} , \\[4mm]
\sin\psi = \dfrac{\ddot{y}}{\sqrt{\ddot{x}^2 + \ddot{y}^2}} = \dfrac{(\ddot{r} - r\dot{\varphi}^2)\sin\varphi + (2\dot{r}\dot{\varphi} + r\ddot{\varphi})\cos\varphi}{\sqrt{(\ddot{r} - r\dot{\varphi}^2)^2 + (2\dot{r}\dot{\varphi} + r\ddot{\varphi})^2}} ,
\end{array}\right\} \tag{38.2}$$

and

$$\tan\psi = \dfrac{\ddot{y}}{\ddot{x}} = \dfrac{(\ddot{r} - r\dot{\varphi}^2)\sin\varphi + (2\dot{r}\dot{\varphi} + r\ddot{\varphi})\cos\varphi}{(\ddot{r} - r\dot{\varphi}^2)\cos\varphi - (2\dot{r}\dot{\varphi} + r\ddot{\varphi})\sin\varphi} . \tag{38.3}$$

39. ELASTIC VIBRATIONS

At the end of Section 37 we saw how the result of our investigation took on a simpler form if β was eventually assumed to be a very small angle. Next we ask what would have become of the investigation if from the outset β had been assumed to be small. If the maximum amplitude β of the vibration is small, then angle φ is small throughout; hence $\sin\varphi$ differs only little from φ, since, as we know,

$$\lim_{\varphi \to 0} \frac{\sin\varphi}{\varphi} = 1 .$$

If, however, we replace $\sin\varphi$ by φ, then our basic assumption (37.1) in Section 37 takes on the form:

$$\frac{l\, d^2\varphi}{d t^2} = -g\varphi .$$

Hence the différential equation becomes

$$\frac{d^2 u}{d t^2} = -\lambda \cdot u . \tag{39.1}$$

It asserts that the acceleration is proportional to the amplitude. If we think, for example, of a vibrating tuning fork whose one end oscillates along the u-axis around $u = 0$ as its position of rest (Fig. 125), it seems reasonable that, the farther it moves away from the rest position, the stronger becomes the elastic force of the fork tending to pull it back. The assumption that this force is but proportional to the elongation is the simplest one can make. So far, however, it seems only qualitatively justified; $\ddot{u} = cu^2$ or $\ddot{u} = c\sqrt{u}$ might be possibilities as well. But we shall again derive mathematical conclusions from our assumption which can be verified quantitatively. Our theory then will cover other physical "models" as well; for example, the motion of a violin string which is plucked or of a point on the vibrating membrane of a drum.

Using the same procedure of integration as in Section 37, we find

$$u' u'' = -\lambda u u' ,$$

$$\frac{d}{dt}\left(\tfrac{1}{2} u'^2\right) = -\lambda \frac{d}{dt}\left(\tfrac{1}{2} u^2\right),$$

$$u'^2 = -\lambda u^2 + c$$

$$[u'(t_0)]^2 = -\lambda[u(t_0)]^2 + c .$$

And if we assume that, with t_0 representing the beginning of the motion, $u'(t_0) = 0$, $u(t_0) = u_0$, then

$$0 = -\lambda u_0^2 + c ,$$

$$u'^2 = -\lambda(u^2 - u_0^2) = \lambda(u_0^2 - u^2),$$

$$\frac{du}{dt} = \sqrt{\lambda}\sqrt{u_0^2 - u^2},$$

$$\frac{dt}{du} = \frac{1}{\sqrt{\lambda}}\frac{1}{\sqrt{u_0^2 + u^2}},$$

$$t = \frac{1}{\sqrt{\lambda}}\int\frac{du}{\sqrt{u_0^2 - u^2}} = \frac{1}{\sqrt{\lambda}}\int\frac{dv}{\sqrt{1 - v^2}}, \qquad v = \frac{u}{u_0},$$

$$t = \frac{1}{\sqrt{\lambda}}\text{ arc sin } v + a ,$$

$$\sqrt{\lambda}(t - a) = \text{arc sin } v ,$$

$$\frac{u}{u_0} = v = \sin[\sqrt{\lambda}(t - a)] = \sin(\sqrt{\lambda}t)\cos(\sqrt{\lambda}a) - \cos(\sqrt{\lambda}t)\sin(\sqrt{\lambda}a).$$

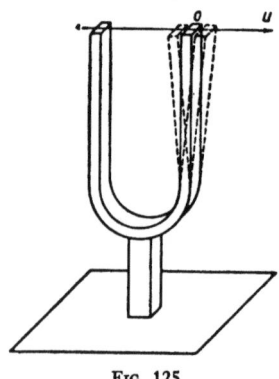

FIG. 125

The function $u(t)$ thus has the form

$$u = [u_0\cos(\sqrt{\lambda}a)]\sin(\sqrt{\lambda}t) - [u_0\sin(\sqrt{\lambda}a)]\cos(\sqrt{\lambda}t)$$

$$= A\sin(\sqrt{\lambda}t) + B\cos(\lambda t) ,$$

where $u_0 = \sqrt{A^2 + B^2}$.

Conversely, we can show by differentiating twice that every expression of the form

$$u = A \sin(\sqrt{\lambda}t) + B \cos(\sqrt{\lambda}t) \qquad (39.2)$$

satisfies equation (38.1); namely,

$$u' = \sqrt{\lambda}A \cos(\sqrt{\lambda}t) - \sqrt{\lambda}B \sin(\sqrt{\lambda}t),$$

$$u'' = -\lambda A \sin(\sqrt{\lambda}t) - \lambda B \cos(\sqrt{\lambda}t) = -\lambda u.$$

Concerning the process of the vibration, statements quite analogous to those in Section 37 can now be made. From

$$\sin(x + 2\pi) = \sin x, \qquad \cos(x + 2\pi) = \cos x$$

we find

$$\sin\left[\sqrt{\lambda}\left(t + \frac{2\pi}{\sqrt{\lambda}}\right)\right] = \sin(\sqrt{\lambda}t), \qquad \cos\left[\sqrt{\lambda}\left(t + \frac{2\pi}{\sqrt{\lambda}}\right)\right] = \cos(\sqrt{\lambda}t),$$

and, consequently, for the function $u(t)$ we obtain from (39.2)

$$u\left(t + \frac{2\pi}{\sqrt{\lambda}}\right) = u(t).$$

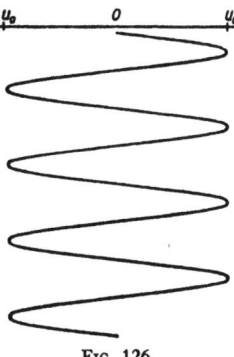

Fig. 126

If again we follow the motion from the moment $t = t_0$, we see that after the time

$$T = \frac{2\pi}{\sqrt{\lambda}} \qquad (39.3)$$

the vibrating point has returned to its initial position and that from there on the motion will simply repeat. T is the period, $\sqrt{\lambda}$ is the "frequency" of the vibration, and $u_0 = u(t_0) = \sqrt{A^2 + B^2}$ is the "amplitude," that is, the maximum elongation.

This result can be tested experimentally. The vibrating tuning fork is placed in front of a projector which throws its greatly enlarged image on a screen. Now, if the tuning fork is rapidly moved in the direction perpendicular to that of its vibration (in Fig. 126, downward), the to-and-fro vibration becomes drawn out into a curve resembling a pure sine curve. This fact and the values of T and the other parameters obtained from the curve prove the correctness of our assumption (39.1).

40. KEPLER'S FIRST TWO LAWS

In investigating the laws of the freely falling body Galileo proceeded from a bold hypothesis. He first tried to reason from the assumption that the nonconstant velocity was proportional to the distance traversed, $ds/dt = c \cdot s$. Finding that this had led him to false conclusions, he assumed instead that nonconstant velocity was proportional to the time of falling, $ds/dt = ct$. That he succeeded with this assumption was due to his brilliant anticipation of the ideas of the calculus applied to a function $f(x) = x^2$ and to a combination of mathematical genius with a genius for experiment. Huygens, too, as we have seen, in investigating the pendulum, proceeded from a mere hypothesis (although one well founded by Galileo's researches) which was subsequently verified by the results.

Newton,[46] in formulating his famous law of gravitation, created the pattern for all subsequent theoretical physical research. His example might be hard to match, but it will continue to be a model long after his law proper has been modified. "Hypotheses non fingo" ("I don't invent hypotheses") are his own words. Mastering analytic geometry and infinitesimal calculus (which he was the first to possess completely), he deduced his law from the facts obtained by Kepler. He showed that, inversely, Kepler's laws were a consequence of the law of gravitation. It is fitting to present here this first and great triumph of differential calculus.

From the encompassing astronomical observations of Tycho Brahe,[47] which excelled all previous ones in accuracy, Kepler,[48] in an ingenious intuition bordering on mysticism and with enormous and tenacious mathematical effort, had, in 1609 and 1623, obtained his three laws of the planetary motions. Before him, all astronomical thought was dominated by the idea of uniform circular motions. Whether astronomers saw the starry firmament revolving uniformly about the earth, and the planets performing circular motions on the revolving firmament, or whether they conceived of the earth and the planets revolving about the sun, they were always thinking in terms of uniform circular motions. Tycho's observational data, although obtained without the aid of telescopes, suggested that the idea of uniform circular motions would have to be abandoned. But it was a long way from this restrictive insight to the realization that Apollonius' theory of conic sections might be applicable to Tycho's observations. By 1609 Kepler had come to the conclusion not that the radius vector from the sun to the planets revolves with uniform angular velocity but that what does change uniformly is the *area* swept over by the radius vector (Fig. 127).

KEPLER'S FIRST LAW. *The radius vector extending from the sun to a planet sweeps out equal areas in equal times.*

Moreover, so Kepler found, the planets do not describe circular orbits around the sun but ellipses.

KEPLER'S SECOND LAW. *The planets describe ellipses around the sun, which stands in one focus of the elliptical orbit.*

These are the facts which Newton took as his starting point. To adapt them to mathematical treatment, he created the concept of an acceleration having a magnitude and a direction; he then inquired what the acceleration is in the case of the planetary motions. Beginning with Kepler's first law, he expressed the area F in polar coordinates, or rather its time derivative dF/dt which, according to the first law, is constant. In Figure 128, dF is the area of a narrow sector of area formed by two closely neighboring radii; it can be approximated by the right triangle SPQ (PQ perpendicular to r at P). The little chip by which it exceeds dF is utterly small in relation to dF, and, as dF itself becomes smaller, this chip diminishes even more rapidly and therefore can be neglected.

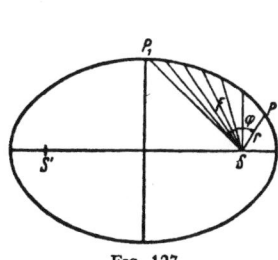

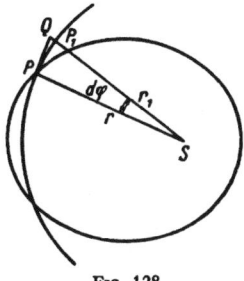

FIG. 127   FIG. 128

Now the area of SPQ is $\frac{1}{2}SP\cdot PQ = \frac{1}{2}rrd\varphi$. In fact, this statement rests on another, although justifiable, approximation: the replacing of the small straight-line segment PQ by the arc PP_1 of the circle. Thus, $dF = \frac{1}{2}r^2d\varphi$; hence

$$\frac{dF}{dt} = \frac{1}{2}r^2\cdot\frac{d\varphi}{dt},$$

and, therefore,

$$\frac{d^2F}{dt^2} = r\dot{r}\dot{\varphi} + \frac{1}{2}r^2\ddot{\varphi}.$$

Kepler's first law, however, asserts that $d^2F/dt^2 = 0$. This means

$$2\dot{r}\dot{\varphi} + r\ddot{\varphi} = 0. \tag{40.1}$$

This is exactly the same expression we found in Section 38 when we transformed the acceleration into polar coordinates; if $2\dot{r}\dot{\varphi} + r\ddot{\varphi} = 0$, then formula (38.3) becomes $\ddot{y}/\ddot{x} = \tan\varphi$. On the other hand, $\ddot{y}/\ddot{x} = \tan\psi$, where ψ gives the direction of the acceleration. Hence, with $\tan\varphi = \tan\psi$, either $\varphi = \psi$ or $\varphi = \psi + \pi$.

This means that the acceleration of the planet is always directed along its straight-line connection with the sun—either toward or away from the sun.*

Newton realized that this follows solely from Kepler's first law and that it holds accordingly for every plane motion in which the radius vector SP describes equal areas in equal times. He showed likewise that the converse of this theorem is true too: if the acceleration of a plane motion is directed toward a fixed point S, then the radius vector SP describes equal areas in equal times.

Indeed, if we choose S as the origin of our coordinate system, it follows from $\ddot{y}/\ddot{x} = \tan \psi = \tan \varphi$ that

$$(\ddot{r} - r\dot{\varphi}^2) \sin \varphi + (2\dot{r}\dot{\varphi} + r\ddot{\varphi}) \cos \varphi$$

$$= \tan \varphi \left\{ (\ddot{r} - r\dot{\varphi}^2) \cos \varphi - (2\dot{r}\dot{\varphi} + r\ddot{\varphi}) \sin \varphi \right\}$$

$$= (\ddot{r} - r\dot{\varphi}^2) \sin \varphi - (2\dot{r}\dot{\varphi} + r\ddot{\varphi}) \frac{\sin^2 \varphi}{\cos \varphi},$$

or

$$(2\dot{r}\dot{\varphi} + r\ddot{\varphi}) \left(\cos \varphi + \frac{\sin^2 \varphi}{\cos \varphi} \right) = 0,$$

or

$$(2\dot{r}\dot{\varphi} + r\ddot{\varphi}) \frac{1}{\cos \varphi} = 0,$$

$$2\dot{r}\dot{\varphi} + r\ddot{\varphi} = 0;$$

that is, $d^2F/dt^2 = 0$, dF/dt is constant, and F is proportional to t.

Newton thus obtained the following general theorem:

THE AREA THEOREM. *A plane motion has its acceleration directed toward a fixed point* S *if and only if the radius vector* SP *covers equal areas in equal times.*

Newton then determined the magnitude of the acceleration from Kepler's second law. To understand it, we shall first transform the equations of the conic sections into polar coordinates.

In analytic geometry the equation of the ellipse usually appears in the form

$$\frac{x^2}{a^2} + \frac{y^2}{b^2} = 1, \tag{40.2}$$

where a is the semimajor, b the semiminor, axis (Fig. 129). Astronomers prefer the equation in the polar form, with one focus S at the origin (Fig. 130):

$$r = \frac{p}{1 + \epsilon \cos \varphi}. \tag{40.3}$$

---

* [In Newton's *Principia* the mathematical treatment of the motions of bodies under given forces in Books I and II is quite separate from its application to planetary motions in Book III, "The System of the World."—EDITOR.]

Here ϵ is the "eccentricity" OS/a, such that $OS = a \cdot \epsilon$. Since

$$BS + BS' = AS + AS' = 2a \quad \text{(string construction!)}$$

and

$$a^2 = b^2 + a^2\epsilon^2, \quad \text{or} \quad b^2 + a^2(1 - \epsilon^2),$$

we obtain from (40.3) for $\varphi = 0$

$$r = SA = a - a\epsilon = \frac{p}{1 + \epsilon};$$

that is,

$$p = (a - a\epsilon)(1 + \epsilon) = a(1 - \epsilon^2).$$

For $\varphi = \pi/2$ we obtain from (40.3)

$$r = SC = \frac{p}{1} = p,$$

which shows that the parameter p is the segment SC.

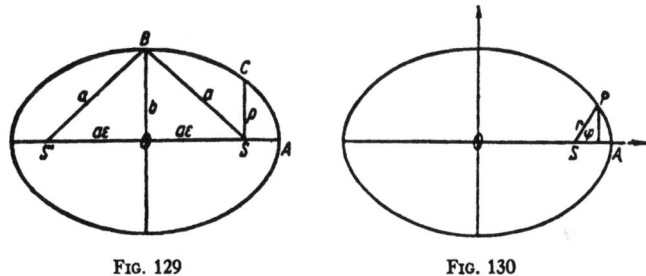

FIG. 129   FIG. 130

The direct transformation of (40.2) into (40.3) is somewhat cumbersome. It is better to obtain (40.3), just as (40.2), from the theory of the conic sections. The polar form (40.3) not only is better suited to the purposes of astronomers but has other merits of its own. For, as we wish to state briefly, the polar form (40.3) gives the other conic sections as well: $\epsilon = 0$ gives $r = p$, that is, the circle; for the ellipse the value of $\epsilon = c/a$ lies necessarily between zero and 1. Thus, while ϵ increases from zero toward 1, a circular-shaped ellipse becomes gradually ever more elongated (Fig. 131). For $\epsilon = 1$ we obtain the parabola (Fig. 132), while for $\epsilon > 1$ we obtain the (left) branch of the hyperbola. The right one is obtained by taking $p < 0$ and $\epsilon < 0$ (Fig. 133).* Thus the value of ϵ determines the nature and shape of the conic section, and similar conics have the same ϵ. All this is stated here without proof.

We now return to our principal task—to determine the magnitude of the

---

* [If we admit negative values of r—which is widely done in analytic geometry—we obtain both branches of the hyperbola by letting φ vary from 0 to 2π.—EDITOR.]

acceleration in a Kepler motion, having previously found its direction. According to Section 38,

$$J^2 = \dot{x}^2 + \dot{y}^2 = (\ddot{r} - r\dot{\varphi}^2)^2 + (2\dot{r}\dot{\varphi} + r\ddot{\varphi})^2 .$$

We saw that $2\dot{r}\dot{\varphi} + r\ddot{\varphi} = 0$ wherever Kepler's first law holds. Hence

$$J^2 = (\ddot{r} - r\dot{\varphi}^2)^2 . \tag{40.4}$$

We know $dF/dt = \frac{1}{2}r^2\dot{\varphi}$, which, according to the first law, is constant; thus, denoting this constant by c,

$$\tfrac{1}{2}r^2\dot{\varphi} = c . \tag{40.5}$$

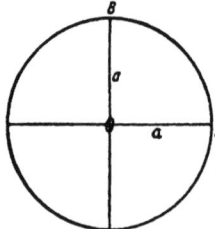

Fig. 131a

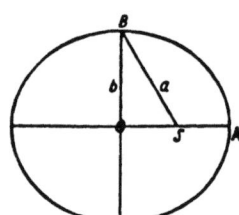

Fig. 131b

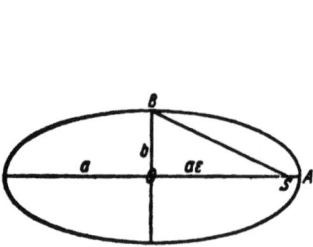

Fig. 131c

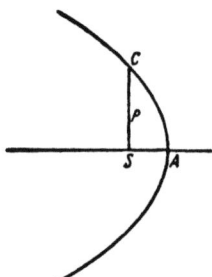

Fig. 132

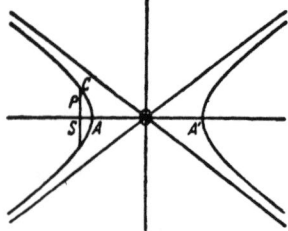

Fig. 133

Further

$$\ddot{r} - r\dot{\varphi}^2 = \ddot{r} - \frac{1}{r^3}(r^2\dot{\varphi})^2 = \ddot{r} - \frac{4}{r^3}(\tfrac{1}{2}r^2\dot{\varphi})^2 = \ddot{r} - \frac{4c^2}{r^3}. \qquad (40.6)$$

On the other hand, according to Kepler's second law, the path of the planet is given by (40.3), which gives

$$r(1 + \epsilon \cos \varphi) = p,$$

or, divided by rp,

$$\frac{1}{p}(1 + \epsilon \cos \varphi) = \frac{1}{r},$$

or

$$\frac{\epsilon \cos \varphi}{p} = \frac{1}{r} - \frac{1}{p}. \qquad (40.7)$$

What we need is $\ddot{r}$. Differentiating (40.7) with respect to t (ϵ and p being constant), we obtain

$$-\frac{\dot{r}}{r^2} = -\frac{\epsilon}{p} \sin \varphi \cdot \dot{\varphi},$$

$$\dot{r} = \frac{\epsilon}{p} \sin \varphi \cdot r^2\dot{\varphi} = \frac{2\epsilon}{p} \sin \varphi \cdot (\tfrac{1}{2}r^2\dot{\varphi}),$$

which, because of (40.5), gives

$$\dot{r} = \frac{2\epsilon c}{p} \sin \varphi.$$

Differentiating a second time with respect to t, we obtain (according to [40.5]) the desired $\ddot{r}$:

$$\ddot{r} = \frac{2\epsilon c}{p} \cos \varphi \cdot \dot{\varphi} = \frac{2\epsilon c}{p} \cos \varphi \cdot \frac{2c}{r^2} = \frac{4\epsilon c^2}{pr^2} \cos \varphi \, ;$$

and, finally, because of (40.6),

$$\ddot{r} - r\dot{\varphi}^2 = \frac{4\epsilon c^2}{pr^2} \cos \varphi - \frac{4c^2}{r^3} = \frac{4c^2}{r^2}\left(\frac{\epsilon}{p} \cos \varphi - \frac{1}{r}\right).$$

This, because of (40.7), means that

$$\ddot{r} - r\dot{\varphi}^2 = -\frac{4c^2}{r^2} \cdot \frac{1}{p}, \qquad (40.8)$$

and, with $p > 0$, we finally obtain, by virtue of (40.4),

$$J = \frac{4c^2}{p} \cdot \frac{1}{r^2}. \qquad (40.9)$$

In (40.8) the right side is negative; hence, according to (38.2) of Section 38, we find

$$\cos \psi = -\cos \varphi, \quad \sin \psi = -\sin \varphi, \quad \psi = \varphi + \pi \, ;$$

that is, the acceleration is directed toward the sun.

Since c and p are constant, the result can be formulated as follows:

NEWTON'S LAW. *The acceleration of a planet which moves according to Kepler's first two laws is always directed toward the sun, and its magnitude is inversely proportional to the square of its distance from the sun.*

But this law does not yet consummate all that Newton discovered, just as it does not draw on all three of Kepler's laws. Before turning to Kepler's third law and to the consequences which Newton derived from it, we shall first follow another of Newton's lines of reasoning. We shall show that, conversely, Newton's law implies Kepler's first two laws; that is, from the statement about the direction and magnitude of the acceleration we shall deduce as a consequence that a planet moves in a conic section according to Kepler's area law. This means that Newton's law is the true core of both of Kepler's first laws. Mathematically, this proof is more difficult than that of the converse.

Before we begin we should first notice that in this section we have assumed only that (40.3) was valid, but we did not assume that $0 \le \epsilon < 1$. Thus (40.3) would still be satisfied if the planet moved in a parabola or on one branch of a hyperbola. It is true that we know of no planet moving in such orbits, although some comets do. However, it shows that we may not expect the mathematical theory which we are about to develop to tell us that the planet moves in an ellipse; it can do no more than show that it moves in a conic section. If, for example, it were to move in a parabola, Newton's law would still be valid.

41. DERIVATION OF KEPLER'S FIRST TWO LAWS FROM NEWTON'S LAW

Motion, according to Newton's law, takes place in space; for Kepler it was a result of observation that the planets move in plane orbits. Hence, if we start with Newton's law, we have to prove first of all that motions do take place in a plane. We therefore think at first of motion as being described by three equations, $x(t)$, $y(t)$, $z(t)$. According to Newton's law, we then have $\sqrt{\ddot{x}^2 + \ddot{y}^2 + \ddot{z}^2} = c/r^2$. Let S be the origin of the coordinate system, $r = SP = \sqrt{x^2 + y^2 + z^2}$, and the acceleration be directed toward S.

How do we express this direction mathematically? We recall how we defined acceleration: it was obtained, as to both magnitude and direction, from the derivatives $\dot{x}$, $\dot{y}$, $\dot{z}$, in the same way as the magnitude and direction of velocity were derived from x, y, z. If the velocity is always directed toward the origin, we have (Fig. 134)

$$dx : dy : dz = x : y : z,$$

or

$$\dot{x} : \dot{y} : \dot{z} = x : y : z. \tag{41.1}$$

Hence for the acceleration to be directed toward S means

$$\ddot{x} : \ddot{y} : \ddot{z} = x : y : z. \tag{41.2}$$

From that follows

$$y\ddot{z} - z\ddot{y} = 0, \quad z\ddot{x} - x\ddot{z} = 0, \quad x\ddot{y} - y\ddot{x} = 0, \qquad (41.3)$$

or

$$\frac{d}{dt}(y\dot{z} - z\dot{y}) = 0, \quad \frac{d}{dt}(z\dot{x} - x\dot{z}) = 0, \quad \frac{d}{dt}(x\dot{y} - y\dot{x}) = 0,$$

and hence

$$y\dot{z} - z\dot{y} = A, \quad z\dot{x} - x\dot{z} = B, \quad x\dot{y} - y\dot{x} = C, \qquad (41.4)$$

where A, B, and C are constants.

At time $t = 0$, point P has a definite position P_0 and a definite initial velocity. By a rotation of the coordinate system we can make the direction of this initial velocity coincide with the z-axis, so that in this new coordinate system $\dot{x}(0) = 0$, and $\dot{y}(0) = 0$. Since the equations (41.4) hold for all values of t, they

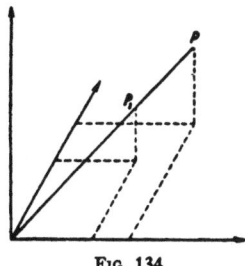

FIG. 134

hold also for $t = 0$. From the third one it follows that $C = 0$ for $t = 0$; and hence, since C is constant, that C is always equal to zero:

$$x\dot{y} - y\dot{x} = 0, \quad \frac{d}{dt}\left(\frac{y}{x}\right) = 0, \quad \frac{y}{x} = \text{Const} = \rho, \quad y = \rho x.$$

This, to be sure, is the equation of a plane through S. We have therefore derived the first result:

> If a planet moves in accordance with Newton's law, it moves in a plane which passes through the sun and which is determined by the direction of the planet's velocity at time t = 0.

Henceforth we may treat the motion again as taking place in a plane. We may rotate the three-dimensional coordinate system so that the plane of the motion coincides with the xy-plane, whose equation is $z = 0$, and thus we may simply disregard z. That is, we have again a plane motion whose acceleration is directed toward S. In Section 40 we learned from Newton that this fact alone implies Kepler's first law, namely, $d^2F/dt^2 = 0$, or, since $dF/dt = \frac{1}{2}r^2\dot{\varphi}$, that $2\dot{r}\dot{\varphi} + r\ddot{\varphi} = 0$, or

$$r^2\dot{\varphi} = 2c; \quad \dot{\varphi} = \frac{2c}{r^2}. \qquad (41.5)$$

Formulas (38.2) of Section 38 then assume again the simple form:

$$\cos \psi = \frac{(\ddot{r} - r\dot{\varphi}^2)}{|\ddot{r} - r\dot{\varphi}^2|} \cos \varphi, \qquad \sin \psi = \frac{(\ddot{r} - r\dot{\varphi}^2)}{|\ddot{r} - r\dot{\varphi}^2|} \sin \varphi.$$

According to Newton's law $\psi = \varphi + \pi$; hence $\ddot{r} - r\dot{\varphi}^2$ must be negative. By (38.1) of Section 38, $|\ddot{r} - r\dot{\varphi}^2|$ is the magnitude of the acceleration. Since, according to Newton's law, this acceleration is inversely proportional to the square of the distance, it follows that

$$\ddot{r} - r\dot{\varphi}^2 = -\frac{\gamma}{r^2}, \qquad (41.6)$$

where γ, the constant of proportionality, is meant to be positive.

Combining (41.5) and (41.6), we obtain

$$\ddot{r} = \frac{4c^2}{r^3} - \frac{\gamma}{r^2}. \qquad (41.7)$$

This much results directly from Newton's law.

If we were next to integrate the differential equation (41.7)—which we can do in accordance with procedures studied before—we would obtain r as a function of t. But that is not our goal. What we wish to determine is the path of the planets; that is—since we have introduced polar coordinates (r, φ)—we wish to determine r as a function of φ. Now both r and φ are functions of t, namely, $r(t)$ and $\varphi(t)$—although not yet known—and hence implicitly r is a function of φ; for, to every value of φ belongs a value of t (the inverse function of $\varphi[t]$), and to this t belongs a value $r(t)$. Thus with any value of φ is paired one value of r, provided $\varphi(t)$ has a one-valued inverse function. This is a point we must first look into; equation (41.5) will tell us what we need.

If the constant $c = 0$, then $\dot{\varphi} = d\varphi/dt = 0$. This means φ is constant and thus has no inverse function. This shows that the point we raised is not self-evident. However, the case $c = 0$ means that the motion is permanently directed toward the sun, and, since the angle does not vary, is rectilinear. It would further mean

$$\ddot{r} = -\frac{\gamma}{r^2}, \qquad\qquad \dot{r}\ddot{r} = -\gamma\frac{\dot{r}}{r^2},$$

$$\tfrac{1}{2}\dot{r}^2 = \tfrac{1}{2}\gamma\frac{1}{r} + \frac{A}{2}, \qquad \left(\frac{dr}{dt}\right)^2 = \frac{\gamma + Ar}{r}, \qquad \frac{dt}{dr} = \sqrt{\frac{r}{Ar + \gamma}}.$$

We have learned from this how to find t as a function of r. The planet course is, anyway, a mere straight line and thus without interest; the planet would simply be falling straight into the sun. The "conic section" has degenerated into an "infinitely narrow ellipse" (Fig. 135) which consists of the straight segment connecting the two foci.

Let us therefore assume $c \neq 0$. For $c > 0$ (in case $c < 0$ we can reason similarly) we see, from (41.5), that $\dot{\varphi}$ is always greater than zero; that is, φ is throughout a monotonic increasing function of t. Hence $\varphi(t)$ has throughout a one-valued inverse function.

We may therefore, without further scruples, regard r as a function of φ and hence transform the differential equation (41.7) for $r(t)$ into one for $r(\varphi)$. According to (41.5), we have

$$\frac{dr}{dt} = \frac{dr}{d\varphi}\frac{d\varphi}{dt} = \frac{dr}{d\varphi}\frac{2c}{r^2},$$

and further

$$\frac{d^2r}{dt^2} = \frac{d}{d\varphi}\left(\frac{dr}{d\varphi}\frac{2c}{r^2}\right)\frac{d\varphi}{dt} = \left[\frac{d^2r}{d\varphi^2}\frac{2c}{r^2} + \left(\frac{dr}{d\varphi}\right)^2\left(-\frac{4c}{r^3}\right)\right]\frac{d\varphi}{dt},$$

and, again, because of (41.5),

$$\frac{d^2r}{dt^2} = -2c\left[-\frac{d^2r}{d\varphi^2}\frac{1}{r^2} + \left(\frac{dr}{d\varphi}\right)^2\frac{2}{r^3}\right]\frac{2c}{r^2} = -\frac{4c^2}{r^2}\left[-\frac{d^2r}{d\varphi^2}\frac{1}{r^2} + \left(\frac{dr}{d\varphi}\right)^2\frac{2}{r^3}\right].$$

FIG. 135

The expression inside the brackets is the derivative of $-(1/r^2)(dr/d\varphi)$ with respect to φ; hence

$$\frac{d^2r}{dt^2} = -\frac{4c^2}{r^2}\frac{d}{d\varphi}\left(-\frac{1}{r^2}\frac{dr}{d\varphi}\right) = -\frac{4c^2}{r^2}\frac{d^2(1/r)}{d\varphi^2}. \qquad (41.8)$$

We have thus expressed the time derivative $\ddot{r}$ in (41.7) in terms of $dr/d\varphi$ and $d^2r/d\varphi^2$, and hence (41.7) assumes the form

$$-\frac{4c^2}{r^2}\frac{d^2(1/r)}{d\varphi^2} = \frac{4c^2}{r^3} - \frac{\gamma}{r^2}$$

or

$$\frac{d^2(1/r)}{d\varphi^2} = -\frac{1}{r} + \frac{\gamma}{4c^2}. \qquad (41.9)$$

This we can write in the form

$$\frac{d^2(1/r - \gamma/4c^2)}{d\varphi^2} = -\left(\frac{1}{r} - \frac{\gamma}{4c^2}\right).$$

Placing $1/r - (\gamma/4c^2) = u$, equation (41.9) takes on the simple form

$$\frac{d^2u}{d\varphi^2} = -u. \qquad (41.10)$$

We recognize this differential equation from Section 39 as the equation of elastic vibration. Accordingly, we find that $u = 1/r - (\gamma/4c^2)$, and hence $1/r$ itself, is a periodic function of φ of the form

$$u = A\cos\varphi + B\sin\varphi.$$

We can write this in a still simpler form. If in a rectangular coordinate system (Fig. 136) we take a point whose coordinates are (A, B), and if (ρ, a) are the polar coordinates of this point, then

$$A = \rho \cos a, \quad B = \rho \sin a. \tag{41.11}$$

Hence we can write

$$u = \rho(\cos a \cos \varphi + \sin a \sin \varphi)$$
$$= \rho \cos (a - \varphi) = \rho \cos (\varphi - a),$$

$$\frac{1}{r} - \frac{\gamma}{4c^2} = \rho \cos (\varphi - a),$$

where a is a constant. Hence

$$\frac{1}{r} = \frac{\gamma}{4c^2} + \rho \cos (\varphi - a),$$

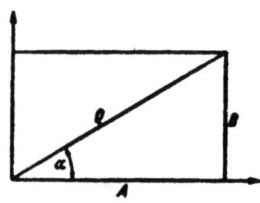

FIG. 136

or, by rotating our coordinate system by a,

$$\frac{1}{r} = \frac{\gamma}{4c^2} + \rho \cos \chi, \qquad (\chi = \varphi - a),$$

$$r = \frac{1}{(\gamma/4c^2) + \rho \cos \chi}. \tag{41.12}$$

In the new (rotated) system of polar coordinates, this is an equation of the form

$$r = \frac{p}{1 + \epsilon \cos \chi} = \frac{1}{(1/p) + (\epsilon/p)\cos \chi},$$

where

$$\frac{1}{p} = \frac{\gamma}{4c^2}, \qquad \frac{\epsilon}{p} = \rho;$$

that is

$$p = \frac{4c^2}{\gamma}, \qquad \epsilon = p\rho = \frac{4c^2\rho}{\gamma}.$$

We have thus proved that the orbit is a conic section, that is, that Kepler's second law, like the first one, follows from Newton's law.

Our procedure in this proof has followed a straightforward plan; only the transformation of (41.9) into (41.10) might be felt as a somewhat artful device. But it led straight to our goal.

42. KEPLER'S THIRD LAW

Kepler's first two laws were proved to be equivalent to Newton's law, according to which the magnitude of the acceleration is

$$J = \frac{\gamma}{r^2},$$

and in which the constant of proportionality γ was found to be

$$\gamma = \frac{4\,c^2}{p},$$

where $c = dF/dt$ is the rate of change of the area, and $p = CS$ is the semi-*latus rectum* of the elliptical orbit, where $p = a(1 - \epsilon^2)$ (Fig. 137). But what is this γ if calculated for the several planets?

Kepler provided the necessary material for this problem too. In 1609 he published his first two laws; on his third law he worked until 1619.[49] He dis-

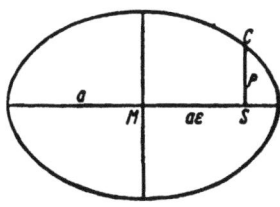

Fig. 137

covered his first two laws by working on the observational data for Mars, whose orbit has—next to Mercury's—the greatest eccentricity ϵ, and where, consequently, he had the best chance of determining the deviation from uniform circular motion. (In fact, Tycho Brahe's observations, on which Kepler based his calculations, referred only to Mars.) The next problem was to compare the motions of the several planets with one another. What motivated him in the search for a relation between them were quasi-mystical ideas derived from Platonism. Plato had imagined each planet to be attached to a massless hollow sphere, with the earth in its center, which performed the diurnal rotation. However, the planet was not fixed on this rotating sphere but revolved on it with uniform circular motion. To an inhabitant of this sphere the planet would thus have appeared as moving uniformly in a circle, whose center and angular velocity were different from that of the sphere. Plato imagined further that the periods of revolution of these spheres, those of the planets relative to their spheres, and the radii of the various spheres had simple rational ratios to each other which ought to be discovered. And the planets, thus circling according to these simple ratios, perceive them as harmonies of infinite beauty vastly more beautiful than those which we enjoy in the musical harmonies with their simple rational proportions.

We must realize that it was from such exalted fancies that Kepler set out to discover his third law, although "not in the manner I had formerly believed, but in a different, yet altogether perfect manner." Indeed, spheres and epicycles had become ellipses, and the relation between the motions of the several planets, which he finally discovered, was this:

KEPLER'S THIRD LAW. *The squares of the periods of revolution of the planets are to each other as the cubes of their major axes.*

The agreement of this formula with the observational material at his disposal was excellent, and Kepler could indeed be satisfied (Table 1).

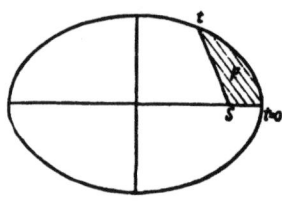

FIG. 138

TABLE 1

Planet	Semimajor Axes* (a)	Period of Revolution† (T)	a^3	T^2
Mercury	0.388	0.241	0.058	0.058
Venus	0.724	0.615	0.379	0.378
Earth	1.000	1.000	1.000	1.000
Mars	1.524	1.881	3.540	3.538
Jupiter	5.200	11.86	140.6	140.7
Saturn	9.510	29.46	860.1	867.9

* With the earth's semimajor axis as unit.
† In years.

What was the significance of this relation for Newton's theory? His constant of proportionality γ had been found to be equal to $4c^2/p$, where $c = dF/dt$, that is, $F = ct + d$, or upon setting $F = 0$ at $t = 0$, $F = ct$ (Fig. 138). After a full period of revolution T, $F = cT$ is the full area of the ellipse, that is, $F = ab\pi$. Hence $ab\pi = cT$, or, according to Section 40,

$$c = \frac{ab\pi}{T} = \frac{a^2(\sqrt{1-\epsilon^2})\pi}{T}.$$

It follows that

$$\gamma = \frac{4c^2}{p} = \frac{4a^4(1-\epsilon^2)\pi^2}{a(1-\epsilon^2)T^2} = 4\pi^2\frac{a^3}{T^2},$$

where $4\pi^2$ is a fixed number, and, according to Kepler's third law, a^3/T^2 has the same value for all planets. Hence the significance for Newton of that law was this:

The factor of proportionality γ in Newton's law has the same value for all planets.

Since Kepler's time the observation of the Jupiter satellites had improved. Galileo had discovered four of them with the newly invented telescope, the four largest of the eleven known at present. Newton calculated their orbits and found that, within the limits of accuracy of the observations, all four not only moved in accordance with Kepler's first two laws but had one and the same value γ, which, however, was much smaller than that of the planets moving around the sun. (Present data show the ratio of Jupiter's γ to the sun's γ to be 1/1,047 for eight of the eleven satellites, and 1/1,053, 1/1,055, and 1/1,058, respectively, for the three others. Newton arrived at the value 1/1,067.)

Utilizing all available astronomical data, Newton treated similarly all the planets which were known to have satellites and obtained the ratio 1/3,121 for the ratio of the γ of Saturn to that of the Sun.

Finally, he calculated it also for the earth. The earth, it is true, has only one moon, so there can be no question of agreement among the γ values of "all moons." But the one moon serves to calculate the value of γ for the earth, since $γ = 4π^2(a^3/T^2)$.

To compare it with the factor γ of the sun, the semimajor axis a and the period of revolution T must, of course, be measured in the previously used units. We took the semimajor axis of the earth's orbit as unit length, and the earth's time of revolution, one year, as the unit of time. For $a = 1$ and $T = 1$ the γ-factor of the sun is $4π^2$. Now the semimajor axis of the moon's orbit is 384,403 km.; that of the earth's orbit 149,504,000 km.; and the period of revolution T of the moon is 27.322 days. (That the length of the "synodic" month—from full moon to full moon—is $29\frac{1}{2}$ days is due to the fact that in the meantime the earth has traveled along on its orbit around the sun and that the moon takes about 2 extra days to get back into the same relative position to sun and earth. But, seen against the fixed stars, the moon revolves around the earth in 27.322 days. It is this "sidereal" month which we need here.) Measured in years, the moon's period of revolution $T = 27.322/365.256$ years.

With these values we compute T^2/a^3 for the moon's motion around the earth:

log 384403	= 5.58478	log 27.322	= 1.43651
log 149504000	= 8.17465	log 365.256	= 2.56260
log a	= 0.41013 − 3	log T	= 0.87391 − 2
log a^3	= 0.23039 − 8	log T^2	= 0.74782 − 3
		log a^3	= 0.23039 − 8
$\dfrac{T^2}{a^3}$	= $3.29 \cdot 10^5$	log $\dfrac{T^2}{a^3}$	= 5.51743

The factor of proportionality for the earth is $γ_e = 4π^2(1/3.29 \cdot 10^5)$.

That shows the factors γ certainly are not proportional to the volumes of the respective celestial bodies. For example, the ratio of the factors γ for Jupiter and the earth is only one-fourth of the ratio of their volumes.

In that connection, Newton discovered an altogether new relation which elevated his law to the stature of a law of nature. He asked himself how the motion of the freely falling body as studied by Galileo compared to the motion of the moon, the former being regarded as another "moon" much closer to the center of the earth. In fact, its distance r from the earth's center is equal to 1 earth radius, while the distance R of the moon is known to be approximately 60 earth radii. The moon's acceleration is γ_e/R^2. If Newton's law were to apply equally to Galileo's freely falling body, its acceleration ought to be equal to γ_e/r^2.

This value we shall next compute and see whether it agrees with Galileo's value for $g = 9.8$. For this purpose we have to convert γ_e into the same units as g, namely, into meters and seconds: T (days) $= T \cdot 24 \cdot 60^2$ seconds: $r = 6.3784 \cdot 10^6 m$, $R = 384.4 \cdot 10^6 m$, which gives $R/r = 60.266$. (These values for r and R are in accordance with recent measurements.)

Substituting these values into the formula

we obtain

$$\frac{\gamma_e}{r^2} = \frac{4\pi^2 \cdot R^3}{T^2} \cdot \frac{1}{r^2} = 4\pi^2 \left(\frac{R}{r}\right)^3 \cdot \frac{r}{T^2},$$

$$\frac{\gamma_e}{r^2} = 9.89 ,$$

while the approximation $R/r = 60.00$ gives

$$\frac{\gamma_e}{r^2} = 9.76 .$$

These values agree with Galileo's $g = 9.8$ as closely as can be expected, considering the accuracy of the underlying measurements.

The bold idea that the falling body is governed by the same law as the cellestial bodies had been conceived by Newton in 1666. At the age of twenty-three he was in possession of the whole theory which enabled him to make the calculation. But with the numerical values then at his disposal his result differed by about 20 per cent from the value of g; it was disastrous. We may suspect that the mistake had come from an inaccurate value of r, the radius of the earth, which was found by measuring the length of a degree of the earth's circumference. Such measurements were already undertaken in antiquity; Eratosthenes carried them out in Egypt according to the same principles which we use today. We measure the distance between two points P_1 and P_2 lying on the same meridian (Fig. 139), then we measure at both points the angle at which the polestar appears, that is, the point of the sky at which the extended axis of the earth is aiming. Then $\varphi_2 - \varphi_1$ is the difference in latitude between P_1 and P_2, that is, the number of degrees of the arc P_1P_2. If, for example, the arc $P_1P_2 = 10°$, and has been measured to be d km. long, then the full meridian of 180° is $18d$ km. long. From there the radius of the earth r is found by dividing $18d$ by π.

We may imagine the suspense with which Newton, in 1682, was awaiting the results of degree measurements which Picard had begun in 1679. Legend has it that, when Newton learned of the corrected radius, he was too excited to insert it into his formulas to see whether now his theory would be verified and that he asked a friend to do it for him. The result was thoroughly satisfactory, and only then did Newton publish his work containing his entire theory.[50]

As mentioned above, Richer observed in 1672 that his pendulum clock, which in Paris had kept perfect time, was losing in Cayenne 2 minutes and 28 seconds

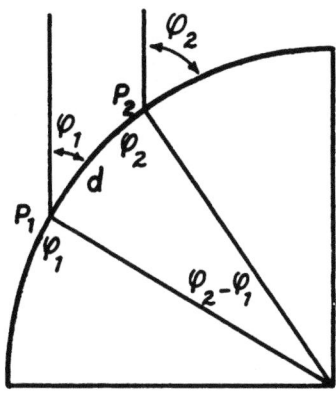

FIG. 139

per day.[51] Long-continued observations and subsequent comparisons with its behavior in Paris, accounting duly for possible temperature variations of the length of the pendulum, left no doubt that the value of g at the equator differed from that in Paris. The accuracy of degree measurements in those days did not yet permit comparison of the circumference of the equator with that of a meridian. Newton had assumed that the earth was flattened at the poles, and Richer's observation confirmed his theory. If a point at sea level on the equator is at a greater distance from the center of the earth than in Paris at latitude 50°, its local acceleration, according to Newton's law, must be smaller in the inverse ratio of the two distances. Hence a pendulum whose period is $T = 2\pi\sqrt{l/g}$ must oscillate more slowly at the equator, just as Richer had observed.

Galileo's discovery that all bodies, regardless of their weight, fall with equal speed, appeared now in an altogether new light. It contained, in fact, the same insight as Kepler's third law, from which Newton had inferred that all planets move around the sun in accordance with the same factor of proportionality γ_s, and all satellites around Jupiter with the same γ_j. In the same way all bodies, the moon included, "fall" toward the earth in accordance with the factor γ_e; at the surface of the earth, that γ becomes Galileo's g, which is constant except for deviations due to the earth's flattened shape. The fact that at a given place all bodies fall with the same speed was to Newton, therefore, an important confirmation of his theory. Since measurements of the speed of falling bodies were rather crude in his time, Newton himself made experiments to assure himself on that point. He constructed a pendulum consisting of a container suspended on a long string, filling it one time with light material, another time with heavy stuff like lead or gold, the geometrical relations being exactly alike in both cases. With the high degree of accuracy characteristic of all pendulum measurements, he convinced himself that the influence of weight on the acceleration at one place amounted to less than 1 per cent—an accuracy comparable to that of Kepler's third law.

After Newton had convinced himself that terrestrial and celestial bodies are subject to the same law, he took a long step further. To appreciate it, a word should first be said about the state of observation of planetary and lunar orbits in Newton's time. Kepler had based his calculations entirely on Tycho Brahe's observations made before telescopes were known, although made with instruments surpassing all earlier ones in size and accuracy of construction. But the first telescopic observations were made already during Kepler's lifetime, and they revealed deviations from his laws of which he had been unaware. And perhaps it was fortunate that he did not know about these deviations because they might have hindered him in discovering his laws. But to Newton it was very clear why things cannot be expected to be altogether simple. For how could Saturn be expected to move in an exact ellipse without having Jupiter's enormous mass, when coming close to Saturn, disturb the ellipse? And how could the moon be expected to describe an exact ellipse around the earth when the gigantic mass of the sun acts upon it and perturbs its orbit around the earth? All these "perturbations" were revealed by the telescope.

All this was in accordance with Newton's law of nature in the generality which he gave to it. For if all bodies, on the earth or in the heavens, impart accelerations to all other bodies in accordance with Newton's law, then each planet acts on every other planet, and on each moon, and the sun acts on all and each, and, finally, though relatively little, they all act upon the sun. Hence it was no longer possible to isolate the sun and some one planet—as Kepler had done—and to act as though there were nothing else in the world. Instead Newton assumed that every celestial body imparts to every other body an acceleration in accordance with his law and its own characteristic gravitational factor γ. But how were all these accelerations imparted, for example, to the moon by the earth, by the sun, by Jupiter, etc., to combine their effects?

Here Newton gave a very plausible hypothesis which was confirmed by his explanations of the "perturbations," the so-called *principle of the superposition of accelerations:* If a body is acted upon by n other bodies $P_1, \ldots, P_n$, each in accordance with their respective factors $\gamma_1, \ldots, \gamma_n$, the total acceleration $\ddot{x}(t), \ddot{y}(t), \ddot{z}(t)$ is given by

$$\ddot{x}(t) = \ddot{x}_1(t) + \ldots + \ddot{x}_n(t) ,$$

$$\ddot{y}(t) = \ddot{y}_1(t) + \ldots + \ddot{y}_n(t) ,$$

$$\ddot{z}(t) = \ddot{z}_1(t) + \ldots + \ddot{z}_n(t) ,$$

that is, equal to the sum of all the several accelerations.

This conception was totally different from Kepler's, even in the case $n = 2$. For his earth's orbit, the sun was fixed, and only the earth was moving. For Newton, even disregarding all other celestial perturbations, both the earth and the sun were in motion: the sun imparts to the earth an acceleration in the direction toward itself of the magnitude γ_s/r^2, while the earth imparts to the sun an acceleration in the direction toward itself of magnitude γ_e/r^2. The

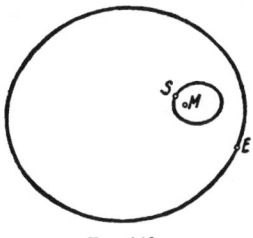

Fig. 140

mathematical treatment of this problem turns out to be much like that of Sections 40 and 41. It yields the result that both bodies describe similar ellipses whose one focus is located in their common center of gravity M (Fig. 140). The elliptical orbit of the sun, however, is only tiny in comparison to that of the earth, because of the relative smallness of the latter's mass. The effect is greatest in the case of Jupiter, where the common center of gravity lies just outside the surface of the sun, as calculated by Newton.

For $n > 2$, however, the mathematical problem becomes exceedingly difficult. Already for $n = 3$, the so-called three-body problem, Newton was unable to find an explicit solution as he had for $n = 2$. To this day that problem has not been solved; we still use approximations, as Newton did. (For Newton, only the system of the fixed stars was stationary; he showed that the center of gravity of the whole solar system is at rest and lies close to the sun because it contains most of the mass of the system.) Newton handled with greatest ingenuity these approximations, which guided him in his work on perturbations and bore out the validity of his two great hypotheses: the principle of accelerations which each body imparts to each other body in accordance with his law and the principle of superposition.

That whole train of ideas, moreover, led Newton to broaden and deepen his inquiry. His point of departure was the problem of the freely falling body. Here the huge earth imparts acceleration to a small falling body. The effect of the entire earth is thought of as concentrated at its center. But why should this be so? Newton was much troubled by this question until he proved this assumption to be correct. For this purpose he divided the whole sphere into thin concentric shells and found that the sum of the accelerating effects of all the shells is equal to that of the whole solid sphere. The "gravitational" factor γ is then the sum of all the γ factors of the shells taken separately (Fig. 141). This involves only the principle of superposition. Next, he treated a single thin shell. Dividing it into small parts, and assuming the same factor γ for all these parts, he proved that each shell imparts to an exterior point an acceleration equal to that imparted by a body located at its center whose γ factor equals that of the whole shell. (The proof requires multiple integration.) Therefore it is indeed

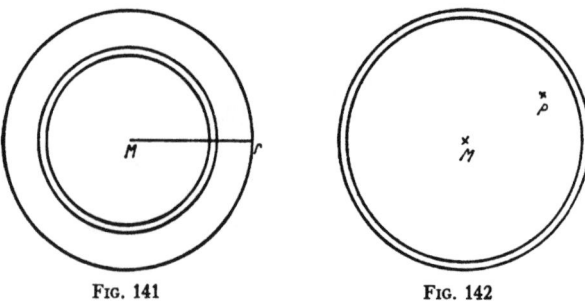

Fig. 141   Fig. 142

correct to replace the total effect of the earth exerted on a point outside or on its surface by that of the center if the latter is assigned the acceleration factor γ_e.

That result justified the reasoning which treated the freely falling body at the surface of the earth as an instance of universal gravitation. Newton drew a further conclusion: If one observes a pendulum in a deep mine shaft, the gravitational effect of the overlying "shells" is eliminated (a shell imparts no acceleration to a point P in its interior [Fig. 142]). Hence γ_e should diminish as one descends into the interior of the earth. That, too, was borne out by experiment.

We are specially interested in the assumption that in this treatment shells of equal volume have the same acceleration factor γ. This assumption grew out of a bold concept. The gravitational effect of the whole earth on the falling body and on the moon's orbital motion could be observed, but the theory of the gravitational effect and the acceleration factor γ of all its several parts was a fiction based on the principle of superposition and stemming from the desire to give to the law of gravitation as general a form as possible. The motion of the freely falling body showed that terrestrial objects of any shape or form are attracted by the earth in accordance with the law of gravitation. But it was a mere hypothesis that any body, no matter of what size or shape, having its specific gravitational factor, imparts an acceleration to any other body, like the sun or the planets. And the idea that the concentric spherical "shells" into which Newton had divided the solid sphere all had the same acceleration factor γ was a further hypothesis which could not be tested in any way.

It was not until 1789 that Cavendish demonstrated with his torsion balance that two lead balls do exert a gravitational effect on each other; and the further experiments by Reich (1838 and 1852), Baily (1842), Cornu and Baille (1873–78), Boys (1894), Braun (1896), and Eötvös (around 1900) have proved that the factor is proportional to the "mass" of these terrestrial bodies, that $\gamma = Gm$. For Newton the statement that the factors γ are proportional to the mass was mere speculation.*

The concept of "mass" implies that of "density" as Mass/Volume—in accordance with the discovery by Archimedes of the specific gravity of bodies—and that of "homogeneity"; a body is called "homogeneous" if it has the same density in all its parts. The previously discussed assumption regarding the spherical shells having the same factor γ may now be formulated by saying that the shells were assumed to be "homogeneous." This, of course, says nothing about the correctness of the assumption of regarding the earth as composed of homogeneous shells.

When we mentioned above that the acceleration factor for Jupiter is only one-fourth as large as we would expect from its volume in comparison to that of the earth, this now means that its density is only one-fourth of that of the earth. The experiments of Cavendish and his successors permitted us to calculate the density of the earth and hence those of the major bodies in the solar system. That of the earth was found to be about 5.5 (approximately that of iron); that of Jupiter, therefore, only about 1.4.

For Newton the idea of "mass" is connected with that of "force." We know from immediate experience that we have to use more muscle power to impart a certain acceleration to a heavy wagon than to a light one and that we hitch two horses before a heavily loaded wagon, while one is enough for a light one. These notions are, of course, rather imprecise and imponderable; effects of friction, inertia, and other influences are intermingled. But they render it plausible that it is not d^2s/dt^2 which is the sole measure of the muscular effort but

* [It may be noted that Toeplitz' whole treatment of the "acceleration factor γ" and his avoidance of "mass" and "force" differ deliberately from Newton's, who places the definitions of "mass" ("quantity of matter") and of "force" at the very beginning of his *Principia*, and who, in formulating his law of gravitation, speaks of the "force" or "power" of gravity by which all bodies attract each other as being in proportion to the "quantity of matter" which the bodies contain. Cavendish and his successors undertook their experiments not to prove that terrestrial bodies attract each other but to measure, with ever higher accuracy, the value of the force between two bodies of known mass and at known distance from each other, that is, to determine the value of the "gravitational constant" G in Newton's law $F = G \cdot [(M_1 \cdot M_2)/r^2]$.—EDITOR.]

$m(d^2s/dt^2)$. It is the product of mass and acceleration which Newton calls "force." Using such an unqualified term, he can speak of a "superposition of forces," of a "parallelogram of forces," and can regard force as the "cause" of motion, or rather of the deviation from rectilinear, uniform motion. But all that does not really explain anything. The "force" which the sun exerts on the earth is for him $m_e(\gamma_s/r^2) = G(m_sm_e/r^2)$; that exerted by the earth on the sun is $m_s(\gamma_e/r^2) = G(m_sm_e/r^2)$, which two are thus equal to each other. This symmetry is mathematically very satisfying and enhances the aesthetic quality of the formulas of celestial mechanics. It states that, of two bodies which attract each other, each exerts the same force on the other. This principle that *actio = reactio* results (in the presentation given here) from the proof that the acceleration factor γ is proportional to the mass.

Newton's investigations went beyond problems of gravitation. He wanted to bring all physical phenomena into the fold of mathematical treatment, not only those covered by his law. He tried magnetic forces, but his "crude attempt," as he calls it, did not lead him to Coulomb's law, which he might have expected to find, but to the tentative placing of the force as equal to γ/r^3. He conjectured about other kinds of "forces" as well and observed that his principle *actio = reactio* seemed to be generally valid in nature. For this reason he placed this principle, together with that of superposition, at the very beginning of his general mechanics, that is, at the beginning of the *Principia*, Book I ("The Motion of Bodies"). This means that he considered only forces for which both principles are valid.

Let us, to resolve any doubt, still apply the concept of "force" as conceived by Newton to some examples which we studied earlier.

1. *Uniform motion constrained to a given path.*—Here we proved the acceleration to be perpendicular to the path and proportional to the curvature. We used as an illustration the case of two railroad trains, one heavy, one light, moving with equal speed around a curve, and we pointed out the greater strain on the rails exerted by the heavy train. This fact made us realize that the "stress" is not given by the acceleration alone. We now see that the concept "force" takes care of this intuitively felt "stress." For "mass times acceleration" is proportional to this "stress," provided this has any tangible meaning to us.

2. *Motion on an inclined plane.*—Here we had

$$\frac{d^2s}{dt^2} = g\cos\alpha, \qquad \frac{ds}{dt} = gt\cos\alpha; \qquad s = \tfrac{1}{2}gt^2\cos\alpha.$$

But we did not yet compute the magnitude and the direction of the acceleration. We shall do this next (Fig. 143). We have

$$x = s\sin\alpha, \qquad y = s\cos\alpha.$$

Hence

$$\dot{x} = gt\cos\alpha\sin\alpha, \qquad \dot{y} = gt\cos^2\alpha;$$

$$\ddot{x} = g\cos\alpha\sin\alpha, \qquad \ddot{y} = g\cos^2\alpha;$$

$$J^2 = \ddot{x}^2 + \ddot{y}^2 = g^2\cos^2\alpha(\cos^2\alpha + \sin^2\alpha) = g^2\cos^2\alpha.$$

Hence
$$J = g \cos a$$
and
$$\cos \psi = \frac{\ddot{x}}{J} = \sin a ; \qquad \sin \psi = \frac{\ddot{y}}{J} = \cos a ;$$
that is,
$$\psi = \frac{\pi}{2} - a ,$$

which means that the acceleration lies in the direction of the path. The acceleration g imparted by the earth is directed vertically downward; but here the inclined plane prevents the free vertical fall. In what way does that produce a "stress" on the plane? The principle of superposition provides an answer, for, in addition to gravity which produces the acceleration,

$$\ddot{x}_1 = 0 , \quad \ddot{y}_1 = g ,$$

there is the "stress" exerted at the surface of the plane which results in the actual motion. Let $\ddot{x}_2$, $\ddot{y}_2$ be the acceleration due to this second force. If the

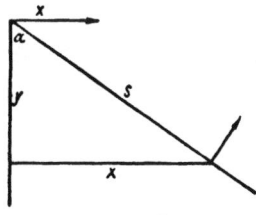

Fig. 143

superposition principle is a universal natural law, it must apply also to the combined action of these two forces. Hence

$$\ddot{x} = \ddot{x}_1 + \ddot{x}_2 , \quad \ddot{y} = \ddot{y}_1 + \ddot{y}_2 ,$$
or
$$g \cos a \sin a = 0 + \ddot{x}_2 , \quad g \cos^2 a = g + \ddot{y}_2 ;$$
hence
$$\ddot{x}_2 = g \cos a \sin a ,$$
$$\ddot{y}_2 = g \cos^2 a - g = -g(1 - \cos^2 a) = -g \sin^2 a .$$

Hence, also,
$$\ddot{x}_2^2 + \ddot{y}_2^2 = g^2 \sin^2 a(\cos^2 a + \sin^2 a) = g^2 \sin^2 a ,$$
or, also,
$$J_2 = g \sin a ,$$
and
$$\cos \psi_2 = \frac{\ddot{x}_2}{J_2} = \cos a , \quad \sin \psi_2 = \frac{\ddot{y}_2}{J_2} = -\sin a , \quad \psi_2 = -a ;$$

that is, the direction of the "stress" is perpendicular to the plane, pointing obliquely upward.

The magnitude of the "stress," on the other hand, is mg sin a, if m is the mass of the falling object; it is thus independent of time and velocity. Only friction, which we have left out of consideration, can cause the stress on the plane to increase with increasing velocity.

All this gives us an idea of the far-reaching significance of the principle of superposition; it leads us to an understanding of the "normal forces" which appear when a body carries out a "constrained" motion rather than moving freely in space. Such constrained motions are of great importance.

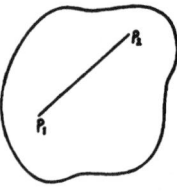

Fig. 144

So far we have considered only single mass points. If we want to free ourselves from that limitation and treat extended bodies, we must begin to consider several, or even infinitely many, mass points which are "rigidly" connected with one another (Fig. 144). For any two of these points, the expression

$$P_1P_2^2 = (x_1 - x_2)^2 + (y_1 - y_2)^2 + (z_1 - z_2)^2$$

must then be independent of time t. That, however, can be regarded as a fixed condition for the coordinates of the several mass points, as though we were requiring that for each of them an equation $f(x, y, z) = 0$ be satisfied; that is, that they should be constrained to move on a certain surface. The "normal forces" are then the intramolecular forces which keep the body "rigid," and these are even more mysterious and less accessible to experiment than those in the fall on an inclined plane. The problem thus arises to treat mathematically the motion of any rigid body solely on the basis of the principles of superposition and of *actio = reactio*.

EXERCISES

In every exercise, easier and more difficult problems are grouped together; the later exercises are in general more difficult and require greater independence.

EXERCISE 1

1. Determine the number $\sqrt[3]{7}$ accurate to three decimal places without using tables.*

2. Prove:

   a) $\sqrt{p}$ is an irrational number, if p is a prime number;

   b) $\sqrt[3]{7}$ is an irrational number.

3. Definition:

   $$n! = 1 \cdot 2 \cdot \ldots \cdot n; \qquad \binom{n}{k} = \frac{n(n-1)\ldots(n-k+1)}{1 \cdot 2 \cdot 3 \ldots k} = \frac{n!}{k!(n-k)!}.$$

   Prove:

   a) $\binom{n}{k} = \binom{n}{n-k}$;

   b) $\binom{n}{k-1} + \binom{n}{k} = \binom{n+1}{k}$;

   c) $\binom{n}{k} - \binom{n-1}{k} = \binom{n-1}{k-1}$.

4. Prove by complete induction that there are $n!$ permutations of n things.

5. Prove by complete induction:

   a) $1^2 + 2^2 + \ldots + n^2 = \frac{1}{6}n(n+1)(2n+1)$;

   b) The binomial theorem

   $$(a+b)^n = a^n + \binom{n}{1}a^{n-1}b + \ldots \binom{n}{k}a^{n-k}b^k + \ldots + \binom{n}{n-1}ab^{n-1} + b^n .$$

6. Prove that a segment divided by a mean proportional is incommensurable to each of its parts. (Explanation: The segment AB is divided by a mean proportional C between A and B if $AC:CB = CB:AB$.)

EXERCISE 2

1. Derive a formula for the sum (closed expression)

   $$1^2 + 3^2 + 5^2 + \ldots + (2n-1)^2 .$$

2. Prove:

$$x + 2x^2 + 3x^3 + \ldots + nx^n = \frac{nx^{n+2} - (n+1)x^{n+1} + x}{(x-1)^2}.$$

3. Suppose $0 < a < b$. Prove that the geometric mean $\sqrt{ab}$ is never greater than the arithmetic mean $(a + b)/2$ and that the harmonic mean

$$\frac{2ab}{a+b} = \frac{1}{\frac{1}{2}[(1/a) + (1/b)]}$$

is never greater than the geometric mean.

4. Prove that for any two numbers x, y it is always true that
   a) $x^2 + xy + y^2 \geq 0$,
   b) $x^4 + x^3y + x^2y^2 + xy^3 + y^4 \geq 0$,
   c) $x^{2n} + x^{2n-1}y + \ldots + xy^{2n-1} + y^{2n} \geq 0$.

5. Let OE be a segment of length 1; at an endpoint let a perpendicular be erected. Let P and Q be two points on this perpendicular such that $EP = PQ$ (Fig. 145). Now let P and Q move off toward infinity upon this perpendicu-

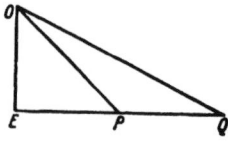

FIG. 145

lar in such a way that $EP = PQ$ always holds. Investigate the value which the ratio $OQ:OP$ approaches. This value is to be obtained by the method of exhaustion.

6. Find a statement whose validity for $n + 1$ can be duly inferred from its validity for n but which is false because there exists no first n for which it holds.

EXERCISE 3

1. Over the segment P_1P_2, which measures one unit in length, an isosceles right triangle is erected with P_1P_2 as one leg. Point P_3 is then chosen so that $P_1P_3 = Q_1P_2$ (Fig. 146). Over P_2P_3 an isosceles right triangle is again erected. P_4 is determined so that $P_2P_4 = Q_2P_3$, etc. Does this infinite process have a concrete result? What is the point which the triangles approach?

2. Let $a > 0$.

   Prove
   $$\lim_{n \to \infty} \sqrt[n]{a} = 1.$$

3. Prove that
   $$\lim_{n \to \infty} (\sqrt{n} - \sqrt{n-1}) = 0.$$

4. Prove the seven rules of calculation for the absolute value given in chapter i, Section 9.

5. Show that the sequences of partial fractions of two different infinite decimal fractions have the same value only when one is an infinite decimal consisting only of 9's after a certain place, and the other is the terminating decimal fraction represented by this infinite decimal.

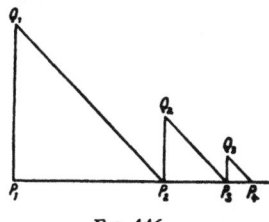

FIG. 146

6. Let $x > 1$, and let

$$s_1 = \frac{2x}{x-1}, \qquad s_2 = \frac{2s_1}{s_1-1}, \qquad \ldots, \qquad s_{n+1} = \frac{2s_n}{s_n-1}, \qquad \ldots.$$

Show that the sequence converges and find $\lim\limits_{n\to\infty} s_n$.

EXERCISE 4

1. Investigate whether the following infinite series converge or diverge:

a) $1 + \frac{1}{4} + \frac{1}{7} + \frac{1}{10} + \cdots + \frac{1}{3n+1} + \cdots$;

b) $\frac{1}{6} + \frac{1}{26} + \cdots + \frac{1}{5^n+1} + \cdots$;

c) $\frac{1}{2} + \frac{1}{8} + \cdots + \frac{1}{3^n-1} + \cdots$;

d) $\frac{1}{1\cdot 3} + \frac{1}{2\cdot 4} + \cdots + \frac{1}{(n-2)n} + \cdots$;

e) $\frac{1}{1\cdot 2\cdot 3} + \frac{1}{2\cdot 3\cdot 4} + \cdots + \frac{1}{n(n+1)(n+2)} + \cdots$;

f) $1 + \frac{1}{2} - \frac{1}{4} + \frac{1}{8} + \frac{1}{16} - \frac{1}{32} + + - \cdots$;

g) $1 - 2^2 + 3 - 4^2 + 5 - 6^2 \pm \cdots$;

h) $\frac{5}{1!} + \frac{5^2}{2!} + \frac{5^3}{3!} + \cdots$;

i) $1 - \frac{2}{3} + \frac{3}{5} - \frac{4}{7} + \frac{5}{9} \mp \cdots$;

j) $1 - \dfrac{\sqrt{2}}{2} + \dfrac{\sqrt[3]{3}}{3} - \dfrac{\sqrt[4]{4}}{4} \pm \ldots \; ;$

k) $1 - \frac{1}{2} + \underbrace{\frac{1}{3} + \frac{1}{4}}_{2} - \underbrace{\frac{1}{5} - \frac{1}{6} - \frac{1}{7}}_{3} + \underbrace{\frac{1}{8} + \frac{1}{9} + \frac{1}{10} + \frac{1}{11}}_{4} \mp \ldots .$

2. *a*) In a triangle another triangle is formed by connecting the three mid-points; in the latter a triangle is formed by connecting its midpoints; etc. (Fig. 147). Do these triangles concentrate about a definite point; if so, about which point?

b) A sequence of circles, not necessarily concentric, is drawn so that each succeeding one is contained in the previous one. The limit of their radii is zero (Fig. 148). Prove that the circles concentrate about a definite point.

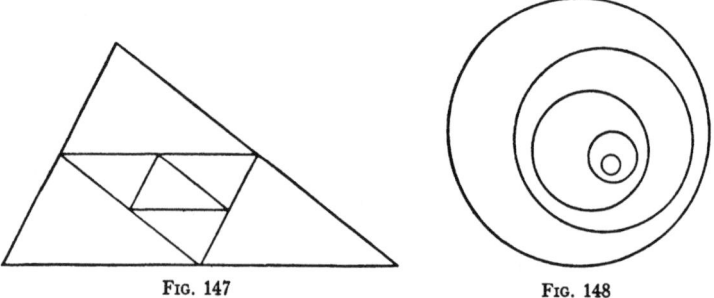

FIG. 147   FIG. 148

3. Prove that

$$\lim_{n \to \infty} n \left(e^{1/n} - 1 \right) = 1 .$$

4. Prove that if the series of positive terms $u_1 + u_2 + \ldots$ converges, then the series $u_1^2 + u_2^2 + \ldots$ also converges.

5. Show by some suitable example of a divergent series that none of the three assumptions of Theorem I in chapter i can be dispensed with.

EXERCISE 5

1. In one diagram draw the graphs of the functions given by

$$y = x, \quad x^2, \quad x^3, \quad x^{-1}, \quad x^{-2}, \quad \sqrt{x}$$

in the interval $-2 \le x \le 2$.

2. Draw the graph of $y = 2^x$;
   a) In the scale $1 = 1$ cm;
   b) In the scale of $1 = \frac{1}{10}$ mm.

3. Find

$$\lim_{n \to \infty} \left[\frac{n^{n+1} + (n+1)}{n^{n+1}} \right]^n .$$

4. For $x > 0$, let

$$x_1 = \sqrt[p]{x \sqrt[q]{x}}, \qquad x_2 = \sqrt[p]{x \sqrt[q]{x x_1}}, \qquad \ldots, \qquad x_n = \sqrt[p]{x \sqrt[q]{x x_{n-1}}}, \qquad \ldots$$

Find a formula which expresses x_n in terms of x and investigate

$$\lim_{n \to \infty} x_n \quad \text{for} \quad pq > 1.$$

5. Construct a series $u_1 + u_2 + \ldots + u_n$ with terms of alternating sign, for which

$$\lim_{n \to \infty} u_n = 0, \quad |u_n| \geq |u_{n+1}|$$

fails infinitely often and yet the series converges.

6. a) Determine the exact sums of the series

$$\frac{1}{1 \cdot 2} + \frac{1}{2 \cdot 3} + \frac{1}{3 \cdot 4} + \quad \text{and} \quad \frac{1}{1 \cdot 2 \cdot 3} + \frac{1}{2 \cdot 3 \cdot 4} + \ldots .$$

b) Find a formula to express the sum

$$\frac{1}{1(x+1)} + \frac{1}{2(x+2)} + \ldots + \frac{1}{n(x+n)}.$$

Show that for fixed x and increasing n this sum tends to a limit, which has a very simple value if x is a positive whole number.

EXERCISE 6

1. a) On the basis of graphical representation, and also by proof, find for what value of x on the relation $\sqrt{x} > \log_2 x$ holds.

   b) In another diagram the graphs of $\log x$, $\sqrt{x}$, $\sqrt[3]{x}$, $\sqrt[4]{x}$ are to be drawn, extending to somewhat large values of x. Investigate the analogous relations $\log x \leq x < \sqrt[3]{x}$, etc.

2. Fermat applied his method (chap. ii, Sec. 12) to $y = 1/x^k$ for $k > 1$ and integral. He calculated the area lying to the right of $x = b > 0$, and bounded by the curve and the x-axis, by allowing x to run through the values b, $b(1 + \delta), b(1 + \delta)^2, \ldots$, calculating the sum T_n belonging to this partition, and, finally, making δ tend to zero. Carry this through!

3. Let $f(x) = 0$ for $x = \frac{1}{2}, \frac{1}{4}, \frac{1}{8} \ldots$. For all other values of x between 0 and 1 let $f(x) = 1$. Does $\int_0^1 f(x)\, dx$ exist, and what value does it have?

4. What is

$$\lim_{n \to \infty} \left(\sqrt{n + \sqrt{n}} - \sqrt{n} \right)?$$

5. Prove that $s_n = 1 + \frac{1}{2} + \frac{1}{3} + \ldots + (1/n) - \log n$ has the properties:
   a) s_n always decreases with increasing n;
   b) s_n is always positive and smaller than 1.

6. If $y = 1/x^s$ and $s > 0$, y becomes infinite as x approaches 0. If $s = p/q$ and $p < q$, then it can be proved by a small modification of Fermat's proof (chap. ii, Sec. 12) that the area to the left of $x = 1$ bounded by the curve and the x-axis and y-axis has nevertheless a finite value. Prove this.

EXERCISE 7

1. Obtain the value of $\int_0^1 e^x dx = e - 1$ from the definition of the definite integral.

2. Differentiate the following functions:

   a) $f(x) = \frac{1}{6}x^7 + \frac{1}{4}x^4 + \frac{1}{3}x^2$;

   b) $x^3(4x^3 - 3x^4 + 15)$;

   c) $(x^2 + a^2)(x^2 - a^2)$;

   d) $(3x + 2)^2(3x - 2) - (3x - 2)^2(3x + 2)$;

   e) $(x^2 + 3)(x^4 - 3x^2 + 9)$;

   f) $\dfrac{x^3 + ax^2 + a^2x + a^3}{x + a}$;

   g) $\dfrac{x^5 - 3x^4 + 9x^3 - 23x^2 + 36}{x^2 + 9}$.

3. Integrate the following functions:

   a) $9x^9 + 7x^6 + 6x^5$;   c) $x^3(x - 1)^2$;   e) $\dfrac{x^5 + 1}{x + 1}$.

   b) $(x^3 - 7)(x^3 + 7)$;   d) $\dfrac{x^4 - a^4}{x^2 + a^2}$;

4. Let $f(x) = x^3 - 3n^2x + 3$, and n be the day of the month of the birthday of the reader. Discuss the curve as to its maxima and minima (i.e., sketch its approximate shape).

5. Form the first derivative of $(x^2 - 1)/2$, the second derivative of $[(x^2 - 1)^2]/(2^2 \cdot 2!)$, the third derivative of $[(x^2 - 1)^3]/(2^3 \cdot 3!)$, etc.

6. Prove the inequality used by Kepler (chap. iii, Sec. 22)

$$\frac{\log a - \log b}{a - b} < \frac{1}{\sqrt{ab}}$$

by comparing $\log x$ with $\sqrt{x} - 1/\sqrt{x}$.

EXERCISE 8

1. Differentiate:

   a) $\dfrac{x^4 + 5}{x^4 - 5}$;   d) $\left(\dfrac{1}{x^2 + 1}\right)^{79}$;

   b) $\dfrac{x^2 + 6x - 91}{x - 7}$;   e) $\dfrac{x^2 + 1}{x}$;

   c) $\dfrac{1}{x + 1} - \dfrac{1}{x - 1}$;   f) $x \log x - x$;

g) $\log[(7x+6)^2]$;   j) $\log(\log x)$;

h) $(\log x)^n$;   k) $\log \dfrac{x^2+1}{x^2-1}$.

i) $(x^3+5)^{19}$;

2. Integrate:

a) $\log x$;   c) $\dfrac{\log x}{x}$;   e) $(5x+7)^{19}$;   g) $\dfrac{2x}{x^2+1}$.

b) $x \log x$;   d) $\log(\log x)$,   f) $\dfrac{1}{(2x-3)^5}$

3. Compute the following definite integrals:

a) $\displaystyle\int_0^1 (2+4x^2)\,dx$,   d) $\displaystyle\int_{-2}^2 x^5\,dx$,   g) $\displaystyle\int_0^2 \dfrac{dx}{(x-x)^2}$.

b) $\displaystyle\int_1^5 \dfrac{dx}{x}$,   e) $\displaystyle\int_{-25,857}^{25,857} x^{23}\,dx$,

c) $\displaystyle\int_{-2}^2 x^{16}\,dx$ f) $\displaystyle\int_{-1}^{-2} \dfrac{dx}{x}$,

4. The "logarithmic derivative" of a function is given by the expression

$$\frac{f'(x)}{f(x)} = \frac{d \log f(x)}{dx}.$$

a) Give the logarithmic derivatives of ax^n and $a(x-c)^n$.

b) Prove that the logarithmic derivative of a product is equal to the sum of the logarithmic derivatives of the factors.

c) Form the logarithmic derivative of $(x-a_1)^{a_1}(x-a_2)^{a_2} \ldots (x-a_p)^{a_p}$.

d) Form a function whose logarithmic derivative is equal to 0, 1, 2x.

5. Form a function which is continuous and monotonic at $x = 0$ but which does not have a differential quotient at $x = 0$.

6. Establish, for the following functions, where they are defined, where they are continuous, and where they are differentiable:

a) $y = \sqrt[3]{x}$;   c) $y = \sqrt{|x^2-4|}$;

b) $y = \sqrt{x^2-6x+8}$;   d) $y = \sqrt[4]{|x+1|}$.

EXERCISE 9

1. By means of the graphs of $y = x$ and $y = \sin x$, prove the inequality used by Aristarchus (chap. i, Sec. 5):

$$\frac{\sin x}{\sin y} < \frac{x}{y} \quad \text{for} \quad 0° < y < x < 90°.$$

2. Differentiate:

a) $e^{(x^2)}$;

e) $x^{\sin x}$;

i) $\arccos \sqrt{1 - x^2}$;

b) $\tan(x^2)$;

f) $\log \sin x$;

j) $\arcsin(1 - x^4)$;

c) $e^{\cos^2 x}$;

g) $\log \dfrac{1}{\sqrt{x}}$;

k) $\log \dfrac{x^2 + 1}{x^2 - 1}$.

d) $e^{(e^x)}$;

h) $\arctan \sqrt{\dfrac{1 - x}{1 + x}}$;

3. Integrate:

a) $\dfrac{x^2 dx}{x^4 - 2x^2 + 1}$;

d) $\sin x \cos x \, dx$;

g) $\sin^2 nx \, dx$;

b) $\dfrac{dx}{x^4 + 5x^2 + 6}$;

e) $\sin nx \cos nx \, dx$;

h) $\tan x \, dx$;

c) $\dfrac{dx}{1 + x + x^2}$;

f) $\cos^2 nx \, dx$;

i) $\dfrac{\cos 3x}{\cos x} \, dx$.

4. a) Form the inverse function and its derivative for

$$y = \sqrt[a]{x^{10}}, \qquad y = \frac{10 - 14x}{35x - 25}.$$

b) For what values of a, b, c, d does $y = (ax + b)/(cx \div d)$ have an inverse function?

5. Prove by complete induction that the nth derivative of a product fg has the form of the binomial theorem, namely,

$$(fg)^{(n)} = fg^{(n)} + \binom{n}{1}f'g^{(n-1)} + \binom{n}{2}f''g^{(n-2)} + \ldots + \binom{n}{n-1}f^{(n-1)}g' + f^{(n)}g .$$

Using this theorem, form all the derivatives of $y = \sin x \cos x$.

6. Find $\int_0^\pi \sin x \, dx$ according to the definition of the definite integral of chapter ii. Use the formula

$$\sin x + \sin(x + h) + \ldots + \sin(x + nh)$$

$$= \frac{\sin[x + (nh/2)] \sin[(n+1)h/2]}{\sin h/2} .$$

EXERCISE 10

1. If $y = (\arcsin x)^2$, prove that the relation

$$(1 - x^2)y^{(n+1)} - (2n - 1)xy^{(n)} - (n - 1)y^{(n-1)} = 0$$

holds for all the higher derivatives ($n \geq 2$).

2. Integrate:

a) $\displaystyle\int \frac{1 + \cos^3 x}{1 + \cos x} \, dx$;

b) $\int \cot x \, dx$;

c) $\int \cos x \, e^x dx$;

$d)$ $\int \sin x\, e^x dx$; $\qquad$ $f)$ $\int \dfrac{1+\sin x}{1+\cos x}\, e^x dx$; $\qquad$ $h)$ $\int \tan^2 x\, dx$;

$e)$ $\int x \tan^2 x\, dx$; $\qquad$ $g)$ $\int \sqrt{\dfrac{1+x}{1-x}}\, dx$; $\qquad$ $i)$ $\int \dfrac{dx}{\cos x}$.

3. Evaluate the definite integrals:

$a)$ $\displaystyle\int_{-\pi/2}^{\pi/2} (\sin x)^{1,153} dx$ $\qquad$ $d)$ $\displaystyle\int_0^a \dfrac{dx}{x^2+a^2}$;

$b)$ $\displaystyle\int_0^{2\pi} \sin px \cos qx\, dx$; $\qquad$ $e)$ $\displaystyle\int_{-1}^1 \sqrt{e^x}\, dx$;

$\quad$ $(p, q$ whole numbers$)$;

$c)$ $\displaystyle\int_0^1 \dfrac{x\, dx}{\sqrt{1-x^2}}$; $\qquad$ $f)$ $\displaystyle\int_0^1 \text{arc} \cos \sqrt{1-x^2}\, dx$.

4. Let $f(x)$ be defined for $x \neq 0$ as $x \sin (1/x)$, and for $x = 0$ let $f(0) = 0$. Does $f(x)$ have a derivative at the point $x = 0$? Decide the same question for $f(x) = x^2 \sin (1/x)$.

5. Show the mathematical-physical impossibility of the functional relation initially conjectured by Galileo for a freely falling body: $ds/dt = cs$.

EXERCISE 11

1. The question whether a body subject to no other forces than air resistance can in time come to a standstill only on account of air resistance amounts mathematically to considering those positive functions $x(t)$, for which

$$\frac{d^2x}{dt^2} = - k^2 \left(\frac{dx}{dt}\right)^2 .$$

What are these functions? [*Editorial note.*—Assume air resistance to be proportional to the square of the velocity.]

2. The motion of a freely falling body, taking into consideration the air resistance, is mathematically given by the differential equation

$$\frac{d^2x}{dt^2} = g - k^2 \left(\frac{dx}{dt}\right)^2 .$$

Determine its solutions. [*Editorial note.*—Same assumption as in the preceding problem.]

3. Let $0 < a < b$, and let

$$a_1 = \frac{a+b}{2} \quad \text{and} \quad b_1 = \frac{2ab}{a+b}$$

be, respectively, the arithmetic and harmonic means of a, b; further let a_2, b_2 be, respectively, the arithmetic and harmonic means of a_1, b_1; in general,

$$a_{n+1} = \frac{a_n + b_n}{2}, \qquad b_{n+1} = \frac{2 a_n b_n}{a_n + b_n} .$$

a) Prove the existence of $\lim a_n = a$, $\lim b_n = \beta$.

b) Prove that $\sqrt[n]{n!}$ becomes infinitely large as n increases and that, more exactly,

$$\lim_{n \to \infty} \frac{\sqrt[n]{n}}{n} = \frac{1}{e}.$$

4. Prove that

$$\frac{2 \cdot 4 \cdot 6 \cdot \ldots \cdot 2n}{1 \cdot 3 \cdot 5 \cdot \ldots (2n-1)} = u_n$$

becomes infinitely large as n increases, and hat, more exactly, $u_n/\sqrt{n}$ has a finite limit.

BIBLIOGRAPHY

WORKS ON THE HISTORY OF MATHEMATICS

The principal German works mentioned in the German edition of Toeplitz are two:

CANTOR, M. *Vorlesungen über die Geschichte der Mathematik.* 4 vols. Leipzig: Teubner, 1907–13.

NEUGEBAUER, O. *Vorlesungen über die Geschichte der antiken mathematischen Wissenschaften.* Berlin: Springer-Verlag, 1934.

BELL, E. T. *The Development of Mathematics.* New York: McGraw-Hill Book Co., 1940.

CAJORI, F. C. *A History of Mathematics.* New York and London: Macmillan Co., 1922.

COHEN, M. R., and DRABKIN, J. F. *A Source Book in Greek Science.* New York: McGraw-Hill Book Co., 1948.

HEATH, T. L. *A History of Greek Mathematics.* Oxford: Clarendon Press, 1921.

NEUGEBAUER, O. *The Exact Sciences in Antiquity.* Providence, R.I.: Brown University Press, 1957.

SARTON, G. *Introduction to the History of Science.* Washington, D.C.: Carnegie Institute, 1931–47.

SMITH, D. E. *History of Mathematics.* 2 vols. Boston: Ginn & Co., 1925.

———. *Source Book in Mathematics.* New York: McGraw-Hill Book Co., 1929.

STRUIK, D. J. *Concise History of Mathematics.* New York: Dover Publications, 1948.

VAN DER WAERDEN, B. L. *Science Awakening.* 2d ed. Groningen: Noordhoff Ltd., 1961.

SPECIAL WORKS ON THE HISTORY OF THE INFINITESIMAL CALCULUS

BOYER, C. B. *The Concepts of the Calculus: A Critical and Historical Discussion of the Derivative and the Integral.* New York: Columbia University Press, 1939.

CAJORI, F. C. *A History of the Conceptions of Limits and Fluxions in Great Britain.* Chicago: Open Court Publishing Co., 1919.

BIBLIOGRAPHICAL NOTES

[Asterisk (*) indicates references added in the English edition.]

1. The passages referring to Zeno are to be found in Aristotle's *Physics* iv. 2, 9, and viii. 8.

2. As to Anaxagoras see H. Diehls and W. Kranz, *Die Fragmente der Vorsokratiker* (Berlin, 1934).

3. A compilation of all the passages in Aristotle which refer to the diagonal of a square is to be found on p. 24 of J. L. Heilberg, "Mathematisches zu Aristoteles," *Abhandlungen zur Geschichte der mathematischen Wissenschaften* No. 18 (Leipzig, 1904), pp. 3–49.

4. The reference to Theaetetus occurs in chap. 38, p. 38, of the translation by O. Apelt (4th ed.; Leipzig, 1923).

5. Euclid editions: *Euclides Opera omnia*, ed. J. L. Heiberg and H. Menge, Vols. I–IV, *Elementa* (Leipzig, 1883–85); T. L. Heath, *The Thirteen Books of Euclid's "Elements*," translated from Heiberg's text, with introduction and commentary, Vols. I–III (Cambridge, 1908). Book x of Euclid's *Elements* contains a complete theory of incommensurable quantities.

6. In the first half of the sixth century A.D. the Cilician author Simplicius wrote a commentary to Aristotle's *Physics*. It contains a passage referring to Hippocrates and Antiphon (see also in the text above for a literal quotation from the since-lost *History of Geometry* by Eudemos of Rhodos, 384–322 B.C.). See also "Der Bericht des Simplicius über die Quadraturen des Antiphon und des Hippokrates," ed. F. Rudio, *Bibliotheca mathematica*, III (3d ed., 1902), 7–62; also *Urkunden zur Geschichte der Mathematik im Altertum*, Heft 1 (Leipzig, 1907); also O. Toeplitz, "Der derzeitige Stand der Forschung in der Geschichte der griechischen Mathematik," *Seminar Berichte* (Bonn and Münster), VI (1935), 4–14.

7. This polemic of Aristotle is treated thoroughly by O. Toeplitz in "Das Verhältnis von Mathematik und Ideenlehre bei Plato," *Quellen und Studien zur Geschichte der Mathematik*, I (Ser. B; 1929), 3–33.

8. *Vietae opera*, ed. Schooten (Leiden, 1646). Compare in particular the mathematical tables, *Canon mathematicus*, of 1579; Stevin, *La Disme enseignant facilement expedier par nombres entiers sans rompuz tous comptes se rencontrans aux affaires des hommes* (Leiden, 1585).

9. The *Opus palatinum de triangulis* was published by Valentinus Otho in Neustadt in 1596. The continuation by Pitiscus of the larger fifteen-place tables of Rhaeticus was published in 1613 under the title *Thesaurus mathematicus*.

10. *Scritti de Leonardo Pisano, matematico del secolo decimoterzo publicati da Bald Boncompagni* (Rome, 1857–62). Concerning the cubic equation of the text, which was proposed to Pisano in the presence of Frederick II of Hohenstaufen, compare the bibliography in M. Cantor, *Vorlesungen über die Geschichte der Mathematik* (4 vols.; Leipzig: Teubner, 1907–13), II, 46.

11. Jordanus Nemorarius is the other important Western mathematician in the first half of the thirteenth century. For particulars about him see Cantor, *op. cit.*, pp. 53 ff.

12. O. Spengler, *Der Untergang des Abendlandes* (2 vols.; Munich, 1923); English edition: *The Decline of the West* (New York: Alfred A. Knopf, 1939).

13. As to the history of the number system see K. Menninger, *Zahlwort und Ziffer* (2d ed.; Göttingen, 1958). *Also see D. E. Smith and L. C. Karpinski, *The Hindu-Arabic Numerals* (Boston: Ginn & Co., 1911).

14. Cf. *Ptolemei Opera quae extant omnia*, ed. J. Heiberg, Vols. I, II (Leipzig, 1898, 1903). A German translation of the *Almagest: Des Claudius Ptolemaeus Handbuch der Astronomie*, translated and with explanatory notes by K. Manitius (Leipzig, 1911–13). *A French translation: *Composition mathématique de Claude Ptolémée*, translated by M. Halma, and with notes by M. Delambre (Paris, 1813).

15. In a commentary on Ptolemy by Theon of Alexandria (*ca.* A.D. 365) a reference is found to the twelve-volume work of Hipparchus, *Concerning the Chords of Circles*, which had also been used by Heron of Alexandria (*ca.* A.D. 60). See *Commentaire de Théon sur le premier livre de la composition mathématique de Ptolémée*, ed. M. Halma (Paris, 1821), I, 110. *Also see A. Rome, "Commentaires de Pappus et de Théon d'Alexandrie sur l'*Almagest*," *Studi e testi* (Vatican City), LXXII (1936), 451.

16. A full discussion of the Greek number concept is found in H. Hasse and H. Scholz, *Die Grundlagenkrisis der griechischen Mathematik* (Charlottenburg, 1928; Leipzig, 1940). See also O. Toeplitz, "Das Verhältnis von Mathematik, und Ideenlehre bei Plato," *Quellen und Studien zur Geschichte der Mathematik*, I (Ser. B; 1929), 3–33.

17. As to editions of Archimedes: *Opera omnia cum Commentarii Eutocii iterum*, ed. J. L. Heiberg, Vols. I–III (Leipzig, 1910–15); T. L. Heath, *The Works of Archimedes*, edited in modern notation with introductory chapters (Cambridge, 1897). The reasoning of Section 5 is found in the treatise κύκλου μέτρησις (*Dimensio circuli*).

18. Al-Bîrûnî (973–1048) opens his trigonometric work *Qânûn* with a theorem, equivalent to the addition theorem, which he designates as a postulate of Archimedes. See C. Schoy, *Die trigonometrischen Lehren des persischen Astronomen Abû'l-Raihân Muh·ibn Ahmad Al Bîrûnî, dargestellt nach Al-Qânûn al-Mas'ûdî*, published after the author's death by J. Ruska and H. Wieleitner (Hanover, 1927). Three proofs referred to as Archimedean are contained in Al-Bîrûnî's *Book of Chords* ("Das Buch der Auffindung der Sehnen im Kreise, von Abû'l-Raihân Muh·el-Bîrûnî"), translated and with commentary by Heinrich Suter in *Bibliotheka mathematica* (Zurich), II (1911), 11–78. See also J. Tropfke, "Archimedes und die Trigonometrie," *Archiv der Geschichte der Mathematik*, X (1928), 432–63.

19. The paper by Jakob Bernoulli bearing this title appeared in 1690 in the *Acta eruditorum* and is reprinted in *Jacobi Bernoulli Basileensis opera* (1744), I, 427–31.

20. See note 17. Also see Archimedes, *Die Quadratur der Parabel, und über das Gleichgewicht ebener Flächen*, translated by A. Czwalina ("Ostwalds Klassiker," No. 203 [1923]). *Also see E. J. Dijksterhuis, *Archimidis* (Copenhagen, 1956), chaps. ix and xi.

21. J. L. Heiberg, *Apollonii Pergaei quae Graece extant cum commentariis antiquis*, Vols. I, II (Leipzig, 1891–93).

22. Archimedes, *Über Spiralen*, translated by A. Czwalina ("Ostwalds Klassiker," No. 201 [1923]).

23. The sum of the first *n* cube numbers is given in the algebra *Al-Fachri* of Abû Bekr Muhammed ibn Alḥusain Alkarchi (*ca*. A.D. 1010). See *Extrait du Fakhrî*, ed. Woepcke (Paris, 1853).

24. The case *n* = 3 is treated in Cavalieri's *Geometria indivisibilibus continuorum nova quadam ratione promota*, published in 1635 in Bologna; revised edition in 1653. The case of *n* > 3 was treated in the *Exercitationes geometricae sex* (1647).

25. Fermat, "Proportionis geometricae in quadrantis infinitis parabolis et hyperbolis usus," *Œuvres de Fermat* (Paris, 1891), I, 255–85.

26. See Gregorius a Santo Vincentio, *Opus geometricum quadraturae circuli et sectionum coni* (Antwerp, 1647). The sixth book treats of the hyperbola.

27. The ἔφοδος was discovered by J. L. Heiberg in a manuscript of the library of the monastery S. Sepulchre in Jerusalem. A German translation by Heiberg appeared in the *Bibliotheka mathematica*, Vol. VII (3d ser.; 1907). *A French translation, "La Méthode relative aux théoremes méchaniques," is included in P. ver Eecke, *op. cit.* *See also T. L. Heath, *The "Method" of Archimedes, Recently Discovered by Heiberg* (Cambridge, 1912); and Enrico Rufini, *Il "Metodo" di Archimedo e le origini dell' analisi infinitesimale nell' antichità* (Rome, 1926). *D. E. Smith, "A Newly Discovered Treatise of Archimedes," *Monist*, XIX (1909), 202.

28. The two axioms of Archimedes appear in the treatise *Kugel und Cylinder* ("Ostwalds Klassiker," No. 202 [1922]).

29. Archimedes, *Über Spiralen* ("Ostwalds Klassiker," No. 201 [1923]).

30. The Euclidean proof appears in Book vi of *Elements* as Theorem 27.

31. See Fermat, *Varia opera of 1679;* also *Methodus ad disquirendum maximum et minimum* (*ca*. 1638).

32. The *Stereometria doliorum* was published in 1615 in Linz. A German translation is by R. Klug, *Neue Stereometie der Fässer* ("Ostwalds Klassiker," No. 165 [1908]). See also Johannes Kepler, *Gesammelte Werke*, Vol. IX (Munich, 1960).

33. *Le Opere di Galileo Galilei* (Florence: Edizione Nationale, 1890–1909). The above-discussed ideas of Galileo are found in the Third Day of the *Discorsi e demostrazioni matematiche intorno a due nuove scienze* (Arcetri, 1638). *English translation: *Dialogues concerning Two New Sciences*, translated by H. Crew and Alfonso de Salvio, with an introduction by Antonio Favoro (London: Macmillan & Co., 1914; 3d ed.; Evanston and Chicago; Northwestern University, 1929).

34. The passage from Stifel's *Arithmetica integra* (Nuremberg, 1544) is quoted in Tropfke, *op. cit.* (3d ed.), II, 209.

35. Joost Bürgi's *Arithmetische und geometrische Progress Tabulen* . . . was published in Prague in 1620. See also E. Voellmy, "Joost Bürgi und die Logarithmen," Supplement 5 of the journal, *Elemente der Mathematik* (Basel) (1948). Napier's *Mirifici logarithmorum canonis descriptio* appeared in Edinburgh in 1614. His introductory text to the logarithm tables appeared in Edinburgh in 1618 under the title *Mirifici logarithmorum canonis constructio*. See also *Collected Works of John Napier* (Edinburgh, 1839).

36. See Mark Napier, *Memoirs of John Napier* (Edinburgh, 1834).

37. Briggs, *Logarithmorum chilias prima* (London, 1617). It contains a thousand fourteen-place logarithms. The *Arithmetica logarithmica* of 1624 contains fourteen-place logarithms of all numbers from 1 to 20,000 and from 90,000 to 100,000.

38. Maestlin's letter to Kepler is found in Kepler's *Gesammelte Werke*, XVII, 297. See also P. Epstein, "Die Logarithmenrechnung bei Kepler," *Zeitschrift für den mathematisch-naturwissenschaftlichen Unterricht*, LX (1924), 142–51. Kepler's tables of logarithms were first published in 1624, in Marburg, under the title *Joannis Kepleri chilias logarithmorum ad totidem numeros rotundos*. Now edited in Kepler, *Gesammelte Werke*, IX (Munich, 1960), 275–426.

39. See Barrow's *Lectiones opticae et geometricae* of 1669. Also John Barrow, *Mathematical Works*, ed. W. Whewell (Cambridge, 1860).

40. The modern definition of the differential is indicated above (p. 106). Leibniz' original conception of the derivative as a quotient of two differentials can be truly understood only from the point of view of his own philosophical ideas, especially his "logic of relations."

41. See Leibniz' mathematical writings, edited by C. J. Gerhardt, Vols. I–VII (1849–63). Most important among them is the essay published in 1684 in the *Acta eruditorum*, "Nova methodus pro maximis et minimis, itemque tangentibus, quae nec fractas nec irrationales quantitates moratur, et singulare pro illis calculi genus" (*op. cit.*, V, 220–26). A German translation contains the selection, edited by G. Kowalewski, *Leibniz über die Analysis des Unendlichen* ("Ostwalds Klassiker," No. 162 [Leipzig, 1908]). Newton's manuscript of 1669 is the "Analysis per aequationes numerò terminorum infinitas." Leibniz, according to his own testimony, was stimulated by Pascal's *Lettres de A. Dettonville contenant quelques unes des inventions de géometrie* (Paris, 1659). There is an extensive literature concerning this priority dispute (see Cantor, *op. cit.*, Vol. III). For a more recent treatment see J. E. Hofmann, "Vom Werden der Leibnizschen Mathematik," *Report of the Mathematics Conference in Tübingen, September 23–27, 1946*, pp. 13–35.

42. A complete edition of Vièta's numerous works was published in Leiden in 1646 (*Vietae opera*, ed. Schooten).

43. Descartes's basic work is *La Géométrie* (Leiden, 1637). See also his *Œuvres*, ed. Adam and Tannery (Paris, 1903).

44. The motion of an obliquely projected body is discussed on the Fourth Day in Galileo's *Discorsi*.

45. See Huygens' *Horologium oscillatorium* (Paris, 1673). Also the *Œuvres complètes de Christiaan Huygens*, published by the Société Hollandaise des Sciences (1888–1937).

46. Newton's *Philosophiae naturalis principia mathematica* was first published in London, 1687. A five-volume, though incomplete, edition of Newton's works appeared between 1779 and 1785; *Opera quae extant omnia*, annotated by Samuel Horsely. *For a recent English edition see *Sir Isaac Newton's (1642–1727) Mathematical Principles of Natural Philosophy and His System of the World*, translated by Andrew Motte in 1729; revised and with a historical and explanatory appendix by Florian Cajori (Berkeley: University of California Press, 1934). As to Newton biographies, see Sir David Brewster, *Memoirs of the Life, Writings, and Discoveries of Sir Isaac Newton* (London, 1885). *Edw. N. de Costo Andrade, *Isaac Newton* (London: M. Parrish, 1950). *Selig Brodetsky, *Sir Isaac Newton: A Brief Account of His Life and Work* (London: Methuen & Co., 1927).

47. See *Tychonis Brahe Dani Opera omnia* (25 vols.; Copenhagen, 1913–29).

48. Kepler's first two laws are contained in the *Astronomia nova*, Vol. III of *Gesammelte Werke*, ed. M. Caspar (Munich, 1937). A German translation by Max Caspar, *Neue Astronomie*, appeared in 1929 (Munich and Berlin).

49. Kepler's third law is contained in his *Harmonice mundi* (Linz, 1619; *Gesammelte Werke*, VI [1940]). There is also a German edition, *Weltharmonik*, translated and annotated by Max Caspar (Munich and Berlin, 1939).

*50. This traditional account of the role of the earth's radius for Newton's theory of gravitation began to be questioned in 1884. It has now been convincingly disproved by Florian Cajori. Pointing to the wide range of values of the earth radius current in Newton's time, Cajori shows that what really delayed Newton was the proof—needed for the numerical comparison between the gravitational force at the surface of the earth and at the orbit of the moon—that a solid sphere acts on a mass outside as though its whole mass were located at its center. See Florian Cajori, "Newton's Twenty Years' Delay in Announcing the Law of Gravitation," in *Sir Isaac Newton: A Bicentenary Evaluation of His Work* (Baltimore: History of Science Society, 1928), pp. 127–90.

51. Richer's observations are cited in Newton's *Principia*.

INDEX